T0358183

Risk, Surprises and Black Swans

Risk, Surprises and Black Swans provides an in-depth analysis of the risk concept with a focus on the critical link between the knowledge, or the lack of knowledge, that risk and probability judgements are based on.

Based on technical scientific research, this book presents a new perspective to help you understand how to assess and manage surprising, extreme events known as 'black swans'. This approach looks beyond the traditional probability-based principles to offer a broader insight into the important aspects of uncertain events, and in doing so explores the ways to manage them.

This book recognises the fundamental issues surrounding risk assessment and risk management to help you to understand and prepare for black swan events.

- Complete with international examples to illustrate ideas and concepts.
- Integrates risk management and resilience-based thinking.
- Suitable for a variety of applications including engineering, finance and security.

Terje Aven has been Professor in Risk Analysis and Risk Management at the University of Stavanger, Norway since 1992. He is a Principal Researcher at the International Research Institute of Stavanger (IRIS). Previously he was also Professor in Risk Analysis at the University of Oslo and the Norwegian University of Science and Technology. He has many years of experience as a risk analyst and consultant in industry. He is the author of many books and papers in the field, covering both fundamental issues as well as practical risk analysis methods. He has led several large research projects in the area of risk which have received strong international participation.

Risk, Surprises and Black Swans

Fundamental ideas and concepts in
risk assessment and
risk management

Terje Aven

Routledge
Taylor & Francis Group

LONDON AND NEW YORK

First published 2014
by Routledge
2 Park Square, Milton Park, Abingdon, Oxon, OX14 4RN

and by Routledge
711 Third Avenue, New York, NY 10017

Routledge is an imprint of the Taylor & Francis Group, an informa business

© 2014 Terje Aven

British Library Cataloguing in Publication Data
A catalogue record for this book is available from the British Library

Library of Congress Cataloging in Publication Data
Aven, Terje.
Risk, surprises and black swans : fundamental ideas and concepts in
risk assessment and risk management / Terje Aven.
pages cm
Includes bibliographical references and index.
1. Risk assessment. 2. Risk management. I. Title.
HD61.A9464 2014
338.5--dc23
2014010956

ISBN: 978-0-415-73506-3 (pbk)
ISBN: 978-1-315-75517-5 (ebk)

Typeset in Sabon
by Saxon Graphics Ltd, Derby

Contents

Figures

Tables

Preface

This book is based on the conviction that assessment and management of risk must break free from sole reliance on the traditional probability-based perspective in order to generate an appropriate understanding of complex risk situations and facilitate adequate risk management decisions. Common perspectives include risk being considered as the expected loss, as the pair of losses and probabilities, or the triplet: What can happen? How likely is that to happen? If it does happen, what are the consequences? The probability-based paradigm provides too narrow an approach to risk and uncertainty assessments, the consequences being that decisions could be seriously misguided, as important aspects of risk and uncertainties are concealed and/or inadequately described. The present book is based on new and broader perspectives, which, in addition to risk descriptions based on probability, require additional characterisations that can provide further insights about knowledge and lack of knowledge, as well as potential surprises/black swans. For example, the (lack of) knowledge dimension captures the fact that probability, used as a measure of uncertainty or degree of belief, is not able to reflect the strength of the knowledge on which the probabilities are based, and does not take account of the fact that the assumptions on which the probabilistic analysis is based could conceal important aspects of uncertainties. The 'surprise' part relates to the fact that surprises may occur relative to the knowledge of the analysts or experts conducting the assessment.

The metaphor and concept of the black swan has gained a lot of attention recently and is a hot topic in many forums that discuss safety and risk. It has also been a focus in the scientific community in the aftermath of Nassib Taleb's *The Black Swan* from 2007. In the present book we explore the type of situations and events that are interesting and important when discussing unforeseen and surprising events, and how to conceptualise these, link them to risk and then confront them (assess and manage). Think of the September 11 event, the Fukushima accident in Japan and the Macondo accident in the Gulf of Mexico. They all came as big surprises relative to current knowledge and beliefs. Hence the knowledge dimension is the key to both understanding and meeting the black swans. It relates either to sensing and adjusting to signals and warnings, transferring experiences and insights and thereby learning, or to scrutinising judgements (including probability judgements) as they are always based on some background knowledge.

In relation to both conceptual aspects and the implications for risk assessment and management, the book sets out to provide new insights by integrating recent scientific work, adding more detail to the issues using relevant examples and including features not covered by other publications. A novel feature of the book is that it not only builds on conceptual aspects of risk and uncertainties, knowledge and surprises from the technical risk analysis community, but also adds insights from the quality discourse, where there is a strong focus on variation and unpredictability. The analysis also draws on social science theories, covering high reliability organisations and resilience, which also address the assessment and management of surprises.

Chapter 1 introduces a set of examples and cases that will be used throughout the book to illustrate ideas, principles and methods. These examples and cases range from personal risk management in relation to pickpocketing to offshore risk management and terrorism risk.

Chapter 2 provides a study of the risk concept and related notions such as risk source, hazard, exposure, vulnerability and resilience. The origin of the term 'risk' is reviewed, as well as the ways in which it is typically used in everyday parlance. From this basis, we move to the professional and scientific sphere and reflect on how the risk concept is defined and understood when used in risk assessment and risk management contexts. A historical and development trend perspective analysis, where some underlying patterns in the way risk has been – and is being – understood today, is also provided. Throughout the analysis, examples and cases are used to clarify and illuminate key points. The chapter relies on basic probability theory as described in Appendix A. For example, the reader needs to be able to distinguish between a frequentist probability and a subjective (knowledge-based, judgemental) probability. Appendix B provides a summary of the main categories of definitions of the risk concept used in professional and scientific settings, as discussed in Chapter 2. The terminology used in the book is summarised in Appendix C.

There is obviously a link between risk and knowledge, uncertainty, variation, surprises and black swans. However, describing this link is not straightforward. Chapter 3 aims to give some perspectives and structures which can be employed to meet this challenge, using recent research on this topic in technical risk environments as well as in the social sciences.

From this basis, Chapters 1 to 3, we are ready to address the issue of how to assess risk (Chapter 4) and how to manage risk (Chapter 5), in line with the ideas developed in the earlier chapters and using the examples and cases introduced in Chapter 1, along with several others, as illustrations. Chapter 4 is concerned with how to assess risk in different decision-making situations, reflecting uncertainties, strength of knowledge and potential surprises and black swans. Assessing risk here covers the identification of hazards/threats/

opportunities, their causes and consequences, making judgements about associated uncertainties and likelihood, and describing risk.

Chapter 5 investigates fundamental risk management principles and strategies, such as the cautionary and precautionary principles, robustness and resilience management, cost–benefit types of analyses, risk-based decision criteria, and the concept of (collective) mindfulness as used in studies of High Reliability Organisations (HROs), covering the five principles: preoccupation with failure, reluctance to simplify, sensitivity to operations, commitment to resilience, and deference to expertise. Key questions addressed are how to deal with deep/severe uncertainties, and how to confront the possibility of surprises (black swans). The chapter also includes some conclusions, with recommendations, on how to best manage risk and the unforeseen.

The idea has been to produce an authoritative, scientifically-based risk book which highlights reflections, for quite a broad readership. The book is mainly intended for professionals in the risk field, including graduate students and researchers. However, it should also be of interest to many others, such as managers, policy makers and business people. The book is conceptually advanced, but at the same time it is quite easy to read. Ideas and principles are highlighted, rather than technicalities. Readers would benefit from a basic knowledge in probability theory and statistics, as well as in risk assessment methods, but the book does not require extensive prior knowledge – the key concepts will be introduced and discussed thoroughly within the text.

The book is about fundamental issues in risk assessment and risk management, and it provides clear guidance in this context. However, it does not prescribe which risk assessment method should be used in different situations. It is not a recipe book. What is covered is the overall thinking process related to the conceptualisation of risk, and how we should assess and manage risk, also taking into account the unforeseen and surprises.

The book is used as a textbook in a graduate course in Risk Assessment and Management at the University of Stavanger. It has been developed as part of two research projects founded by the Norwegian Research Council (as part of the Petromaks 2 programme), Gassco, Norwegian Oil and Gas Association and ConocoPhillips. Their support is gratefully acknowledged.

I would also like to acknowledge Enrico Zio, Ortwin Renn, Seth Guikema, Roger Flage, Eirik B. Abrahamsen and Bodil S. Krohn for their valuable input to this book through their collaboration on these projects and relevant papers. This book project would not have become a reality without the initiatives, inspiration and enthusiasm shown by these scholars.

1 Introduction

This chapter presents some examples and cases that will be used throughout the book to illustrate the discussion about risk. The first example relates to pickpocketing and shows how the risk concept is used in such a daily-life situation. It also points to many fundamental issues of risk assessment and risk management – for example, how to determine what is acceptable risk and balance different concerns. The second example addresses a challenge that many people face: delivering a talk in front of a professional audience. The example illustrates the need for seeing the risks in relation to performance: one issue is the possibility of a bad outcome for the speaker. However, equally and probably more important is the goal of giving a very successful talk. Key means for obtaining a desirable outcome will be discussed. The third example points to the need to provide understandable interpretations of the risk and probability numbers produced by risk assessments. Such interpretations are often lacking, the result being poor risk communication, as demonstrated by this example. Also discussed is the common practice in industrial settings of describing risk through probabilities. It is argued that the strength of knowledge that these computed probabilities are based on also needs to be taken into account. The fourth example, which looks into the *Deepwater Horizon* accident in 2010, is used to illustrate the link between risk and unforeseen combinations of conditions and events, which characterise major accidents. Then some reflections are provided on the 2011 Fukushima Daiichi nuclear disaster in Japan. The accident demonstrates that events occur despite the fact that the risks are judged to be minor and acceptable. The sixth example relates to terrorism. Some reflections are made on the role of probabilities in describing risk. The final example relates to the concept of black swans, a term which is now commonly referred to in relation to the unforeseen occurrence of extreme events. The text addresses the issues of defining black swans, and, having done that, discusses how to identify and deal with them.

1.1 Pickpocketing on the Paris Metro

I was on my way back to Norway, buying a ticket to travel from Paris Gare de Lyon to Charles de Gaulle Airport. I had two pieces of luggage, and my wallet was in my left front pocket. Although it was filled with euro, I needed to use my visa card to get a ticket from the machine. I noticed a man behind me looking over my shoulder and I adjusted my position, ensuring that he could not see my code. It was rush hour, Friday evening, about six o'clock and extremely crowded. I did not give any more thought to this over-the-shoulder event; I certainly did not reflect on risks. My focus was getting onto the train. About twenty people were desperately trying to board the train where I was standing; there could hardly be space for even half of them inside the doors. I was among the first to board; all around me people were pushing. Then it happened. I felt an enormous pressure; somebody forced their way on to the train and pushed me and the people close to me so that it was difficult to breathe. My hands had to leave my luggage and my pockets. A few seconds later things calmed down a little. Then I noticed that my wallet was not there. Sweating, I looked around, but I quickly understood that it would be impossible to identify the thief. I had to accept that my wallet was gone. Three hundred euro and several credit cards were lost. Fortunately these cards were not misused, but the event was really stressful. When thinking about it today – several years later – I still experience strong negative feelings. It was a nasty experience.

Now, let us examine this event through the glasses of risk conceptualisation, risk assessment and risk management. Before entering the station to travel to the airport, we can obviously talk about risk related to pickpocketing. However, from my perspective, I was not thinking about this possible event at all before it happened. It came as a surprise to me. Of course, I was aware of the fact that pickpocketing occurs in such places, but I did not think about it during this particular trip. My mind was filled with other thoughts. Nonetheless, it seems natural to say that I faced risk. Using the term 'risk' in this way seems to indicate that risk exists independently of me; it has an objective status. On the other hand, however, it seems impossible to measure this risk without performing subjective judgements. Different people would judge the risk differently. Professional risk analysts may collect data concerning the occurrence of pickpocketing, but also a number of judgements need to be made: which periods of time should be included, what type of events, which stations, etc. The issue is the extent to which these data are representative of the situation I experienced, also taking into account specific circumstances related to my case.

Such measurements could obviously be of relevance for the authorities in supporting decision-making on how to run the trains safely and demonstrate the need for risk-reducing measures, but they do not seem so meaningful for me, the individual person exposed to the risk. I should be able to cope with the problem of pickpocketing without risk numbers. However, further reflection reveals that being informed about the magnitude of the pickpocketing problem in this area during rush hours could provide useful information for me to plan future train journeys. A natural question to ask is the extent to which the risk represented by the historical data on pickpocketing captures the risk the individual person is exposed to. I could introduce a number of risk-reducing measures, but they would not necessarily be reflected by the risk represented by the historical data. For example, I may decide to never board a train when there are more than (say) ten people trying to do this, or decide to avoid getting on the train when I have observed people that seem suspect. I could think about cautionary measures to reduce the consequences if I were unlucky and got robbed – for example, restricting the amount of money in my pocket to just a few euro and keeping all credit cards and the rest of the cash in a safe place in my suitcase.

I have been at that station many times and I will continue to use it. Reviewing all these occasions, statistical data can be produced showing some variation with respect to different factors, such as the number of pieces of luggage and their form, and how crowded the station is. A key question that I need to ask myself is: what level of variation in these factors should I consider to be normal or common? My first reaction after the event was that a crowd level such as that on the day of the robbery should not be seen as common, and special measures would be needed to avoid such peaks of overcrowding. However, later I thought that maybe I should allow such peak periods to be part of the normal variation – since I may need to take the train at such points in time – and instead think of ways to change my management of the risk, implementing a set of measures as indicated in the previous paragraph.

These are some of the conceptual issues we will address in the coming chapters. The meaning of key concepts such as risk, surprise, knowledge and variation will be thoroughly studied. The degree to which receiving and obtaining more knowledge about the phenomenon (pickpocketing) would change the risk will be questioned. Is the risk independent of the information/knowledge? It seems obvious that the measurement of risk is influenced by the information/knowledge, but is this measurement of risk the same as risk? Does some risk exist per se, which is not influenced by the measurements? And how do these types of reflections affect the way we manage the risk? Theoretical analysis has a value in itself, but the foundational ideas about risk addressed here also have implications for decision-making, as we have seen from the above considerations in this pickpocketing example.

1.2 Delivering a talk to an audience of scientists

The time of the talk is determined, and the speaker is planning its execution. Let us call this person John. His long-term goal is to give brilliant talks, in the sense that the audience listens with great interest to what he says and enjoys the way he communicates his message. In addition, for him success requires having a good feeling throughout the whole talk, a feeling characterised by a high level of confidence and "having the audience in the palm of his hand".

To obtain the desired outcome, John implements some types of risk and performance thinking in line with the ideas in this book. The first one relates to the concept of risk and how it is understood. The talk can have many different outcomes, and before it is executed we do not know which one will occur. This is risk, and John is especially concerned about undesirable scenarios – for example, situations that make him feel incompetent. To assess the magnitude of the risk, he needs to introduce a measure (interpreted in a wide sense) of the uncertainties. John is most used to probability and thinks in accordance with this measure, but he is not assigning it at this stage. Rather, he thinks about the means he believes are necessary to ensure the desired results. A key factor is the process he has adopted recently for early preparation of the slides and carrying out a number of trial-talks with a critical audience of colleagues who provide feedback on his performance. These trial-talks can be viewed as elements of a continuous improvement process (plan, do, study, act). He also focuses on several other means to be successful:

(i) Preoccupation with failure

John is focused on failures that could occur – for example, the audience is bored (one person or more), the arguments used in the talk are not valid, the slides are confusing, the message is not clear, and so on. To a large extent risk is about the occurrence of such events – deviations, catastrophes, not meeting aims etc. – and the identification of them is a basic step in any risk assessment. To be able to give a successful talk, the list of potential failures is studied and a check is made that the means implemented are sufficient to avoid them. If not, additional measures are required. For example, the trial-talks should give a clear indication as to whether or not the slides are confusing.

Equally as important as the focus on failures is the preoccupation with early signals of failure – for example, if some members of the audience show signs of not listening, closing their eyes and so on. This focus on early signals and warnings is important in relation to risk, as signals and warnings are closely linked to the uncertainties and the knowledge dimensions of risk.

(ii) Reluctance to simplify

Following this principle, we will not allow judgements of risk to be based only on the result of the quantitative expression of risk, such as simple risk matrices showing probabilities of failures and expected losses given failures. Such a risk description on the basis of earlier talks may indicate that the risk that a person known for cavilling behaviour will be in the audience is small and negligible. Reluctance to simplify means that we should not base the judgement of risk only on such simple tools. Another example relates to the reliance on simple rules of thumb: for example, the idea that to ensure a successful talk it is sufficient that one smiles and has a good time on the stage, or that being knowledgeable is sufficient. Reluctance to simplify acknowledges the need for seeing beyond such rules. Such rules – 'truths' and assumptions – may lead to surprises. A 'complete' risk picture is sought, covering not only probabilities but also the 'potential for surprises'.

(iii) Sensitivity to operations

The key here is to be sensitive to what is happening during the talk. For example, if some members of the audience show indications of being bored, actions may be taken, such as changing the focus to a topic one knows always gives a good, immediate response. Getting signals of something that might threaten the success of the talk, thereby increasing uncertainties and risks, requires compensating measures. During the talk information is continuously gathered and the risk description is updated, and proper risk management calls for measures. The risk is monitored during the talk, meaning that John is sensitive to everything that happens. If one is to be able to manage unforeseen events occurring during a talk adequately, not only is a lot of training required – there is also a clear need for preparation, as expressed by the next characteristic (iv).

(iv) Commitment to resilience

This principle concerns the ability to meet unforeseen events and surprises – for example, questions from the audience that the speaker has not thought about, or the fact that a person renowned for cavilling behaviour is among the audience. John needs to think about ways to meet such situations. One approach is to establish a general procedure, which is first based on general reflection, and then deferring the answer to an interval or referring to other experts. Being resilient requires a lot of work and training, and it is obviously very important in order to succeed. Resilience is a well-known principle in risk management to meet threats and uncertainties. Characteristic (iv) represents a commitment to highlight such thinking.

(v) Deference to expertise

This principle could be demonstrated by not trying to answer questions out of one's competence area, but rather pointing to other experts who have the necessary knowledge to give an adequate response.

These five principles constitute the key elements of the concept of (collective) mindfulness as interpreted in studies of so-called High Reliability Organisations (HROs). We will study them in more detail in Section 5.2. The concept of mindfulness concerns awareness and sensitivity to discerning the details which are important for obtaining a high level of performance and avoiding catastrophes. When delivering a talk like this, such awareness is critical; signals indicating failures need to be recognised and adjusted for, measures need to be ready for use when special situations occur, and so on. We will show that the concept of (collective) mindfulness can be nicely rooted in the way we understand risk, which focuses on the risk sources, the signals and warnings, the failures and deviations, uncertainties, probabilities, knowledge and surprises.

1.3 Expressing and communicating the result of a risk assessment

This example goes back many years to the time when I worked in the oil and gas industry. On one particular day I was challenged by one of the top managers of the oil company where I was working. We had conducted a risk assessment of an offshore installation, as a basis for making judgements about the risk level and identifying those areas where one could most effectively reduce risk. I had an active role in the analysis, and on the day in question I presented the results of the assessment to the top management of the company. Then one of the managers asked me: "What is the meaning of the risk and probability figures you have presented?" He referred, for example, to a figure where risk was described by a histogram of loss of life categories (1–2 fatalities, 3–5 fatalities, ... more than 100 fatalities) and associated probabilities. It was a good question. The company planned to spend large sums on the basis of this assessment, and to adequately use its results he needed to understand what the numbers really expressed. The problem was that I could not give the manager a good answer. To this day I fret and grieve over the answer, and I realised that I could not live with this; I am an expert on risk and probability, but I had no clear and convincing answers to this basic question. The example demonstrates that in a practical setting we need to communicate the meaning of the risks and probabilities, and we often fail to do this. This incident is a historical example, but the problem is still relevant. If the manager's question were repeated today for a similar type of assessment, many risk analysts would face similar problems (Aven 2008, p. 146).

The second part of this example relates to the way risk was described in the risk assessment. As indicated above, the risk was presented through probabilities related to various loss of life categories. The numbers produced were derived from extensive calculations using various models and based on a number of assumptions related to the operation and management of the installation – for example:

- The platform jacket structure will withstand ship collision energy of 14 megajoules.
- There will be no hot work on the platform.
- The work permit system is adhered to.
- The reliability of the blowdown system is p.
- There will be N crane lifts per year.

The computed probabilities are thus to be viewed as conditioned on these assumptions. It is common to say that the probabilities are based on some specific background knowledge – which we will refer to as K. Clearly, the results need to be seen in relation to this knowledge since varying K would change the risk numbers. However, the practice in the industry was to use these numbers more or less mechanically to make judgements about whether or not the risk was acceptable. If the probability were below some defined threshold value risks were judged to be acceptable, whereas if the probability exceeded the threshold value, the risk was considered unacceptable and risk-reducing measures were put in place.

How can this practice be justified? Is risk not also related to these assumptions and the related background knowledge? How can we be sure that all procedures will be followed – for example, that the work permit system will be adhered to? Obviously there are uncertainties about the fulfilment of these assumptions, but does this not mean that there are also risk aspects related to these assumptions?

Think about two risk assessments, one which is based on thorough modelling and the best experts available, the other based on a rather crude analysis using personnel with quite weak competence in the field. In theory these two assessments could produce the same risk description if it is restricted to probabilities, as in the above example. Surely it is also necessary to look at the background knowledge of the people carrying out the assessment, and the strength of this knowledge, when using such an assessment in decision-making. In later chapters we will discuss these issues in more detail. We will give further substance to the arguments that the background knowledge may conceal important aspects of risk, and that the risk description and evaluation needs to capture this dimension of risk to properly communicate risk and adequately use risk assessments in decision-making contexts, for example related to judgements about risk acceptability. The thesis is that there is a need for seeing beyond probability to inform the decision-maker in a proper way. Again the example used goes back some years in time, but it is still highly relevant (Aven 2011b). We refer to this case as the QRA (Quantified Risk Assessment) example. It is applicable to a large class of technological systems, including nuclear power plants, oil and gas installations and Liquefied Natural Gas (LNG) plants.

1.4 The Macondo accident

On 20 April 2010, a well control incident resulted in the escape of hydrocarbons from BP's Macondo well onto Transocean's *Deepwater Horizon*, leading to explosions, a fire and the loss of eleven lives. After thirty-six hours of fire, the rig sank. For eighty-seven days hydrocarbons continued to flow from the reservoir through the wellbore and the blowout preventer (BOP), resulting in a gigantic offshore oil spill.

The accident did not involve any new phenomena and processes. Any risk assessment of the activity would identify a blowout as a serious hazard that should be followed up with proper risk management. However, if we look more closely at the explanations for the occurrence of the event, we notice that there are many elements of surprise present (as there are for all types of man-machine disasters; see e.g. Turner and Pidgeon 1997). The main one relates to the fact that we experienced a set of conditions and events, which, viewed in advance, had to be assessed as being extremely unlikely to occur together. This set includes (PSA-N 2013b):

- Erroneous assessments of the results of pressure tests.
- Failure to identify that the formation fluid had penetrated the well, in spite of the fact that log data showed that this was the case.
- The diverter system was unable to divert gas.
- The cutting valve (Blind Shear Ram, BSR) in the blowout preventer (BOP) did not seal the well.

Risk assessments either did not capture such combinations, or if they did, the events would typically be disregarded due to negligible assigned probabilities/risk contributions. The co-chair of the National Commission on the BP *Deepwater Horizon* Oil Spill and Offshore Drilling disaster, Senator Bob Graham, stated in his remarks that "the *Deepwater Horizon* disaster did not have to happen. It was both preventable and foreseeable" (HT Politics 2011). By using these words, the senator probably wished to express his opinion that such an event was fairly likely to occur given the existing safety regime and culture. Of course, the specific accident scenario that occurred was not foreseeable. However, although it was unforeseen, it was not unthinkable or unimaginable. In this book we will study in more detail how risk relates to unforeseen events and surprises, the motivation being to improve our ability to manage the risk related to the occurrence of such sequences of conditions and events, and hopefully to avoid the disasters.

1.5 The Fukushima accident

Let us consider the Fukushima Daiichi nuclear disaster in Japan in March 2011. Aven (2013a) reports statements from risk analysts: "Until this event, no one had conceived it a possibility that a tsunami would simultaneously destroy all back-up systems as well as prevent outside support from reaching the site." This statement sounds somewhat strange in view of the investigation committee, which concluded that the government and the operator TEPCO failed to prevent the disaster, not because a large tsunami was unanticipated but because they were reluctant to invest time, effort and money in protecting against a natural disaster which was considered unlikely (Yamaguchi 2012). In other words, the risk was found to be acceptable; the utility and regulatory bodies were overly confident that events beyond the scope of their assumptions would not occur (Wallace 2012). Hence, the event came as a surprise for many people, although it was not unforeseen or unthinkable in the strict sense of the words.

This example is a point of departure for a later discussion in Section 5.1 regarding the basis for making judgements about risk acceptability. A key issue is the extent to which, in a practical risk management context, we can ignore risk that has assigned probabilities which are small or negligible.

1.6 Terrorism risk

For intentional acts (e.g. terrorist attacks), the risk description needs to be much more dynamic than for safety issues. An event occurring on the other side of the world could quickly change the risk assessment of such acts, and the probabilities as well as the knowledge that these probabilities are based on. The same could be said about the result of surveillance and intelligence work. Clearly, being sensitive to signals and warnings of attacks is essential, since avoiding attacks is of course to be preferred in comparison with relying on the ability of effective barriers to reduce the consequences of the attacks. Robustness and resilience are always warranted, but investment and efforts have to be carefully balanced against costs and other values that are valued in a society, such as openness and freedom of movement.

The Norwegian Police Security Service (PST) has defined four categories of threat levels (PST 2013):

(1) Low: One or more parties may have the intention, but are not thought to have the capacity, to strike at specific interests.
(2) Moderate: One or more parties may have the intention and capacity to strike at specific interests.
(3) High: One or more parties have the intention and capacity to strike at specific interests. There is an unspecified threat.
(4) Extreme: One or more parties have the intention to strike at specific interests. There is a specific threat. No further warnings are to be expected before a strike is carried out.

Similar systems exist in other countries – for example, the UK government uses five categories as follows: low – an attack is unlikely; moderate – an attack is possible but not likely; substantial – an attack is a strong possibility; severe – an attack is highly likely; and critical – an attack is expected imminently (UK Gov 2014).

The UK levels refer to likelihood but without any reference to the quantitative scale from 0 to 1 normally used for probabilities. In the PST system (as in the UK system) the threat level categories are linked to some conditions concerning parties' capacities and intentions to strike at particular interests. These examples are not uncommon in the security community. There is scepticism about the use of probabilities in the normal sense. For example, this issue is discussed by Littlewood and Strigini (2004); see also Littlewood *et al.* (1993). These authors point to some of the problems raised:

(a) Security failures are deliberate and thus not open to probabilistic analysis and modelling. The attackers know what they are doing, so where are the uncertainties?
(b) Probability is difficult to use because of the essentially unrepeatable nature of the key events.

As commented by Littlewood and Strigini (2004), the system owner and the defenders will not normally be in possession of knowledge about when the attacker will act and in what way; there are uncertainties. As probability is a tool for representing or expressing uncertainties, it also enters the scene in such contexts. However, it is not clear how we should interpret such probabilities, and how to use them in this setting. The events considered are unique, thus excluding probabilities which are based on a frequentist interpretation. The alternative is to employ subjective (also referred to as judgemental or knowledge-based) probabilities (see Appendix A), but many researchers and security people are sceptical about the use of this kind of probability. A well-known problem with specifying (subjective) probabilities in security contexts is that they are critically linked to risk management responses (see e.g. Brown and Cox 2011): for example, the analysts may assign a high probability number for an attack against specific facilities, with the result that some protective measures are implemented. However, this action may cause potential attackers not to consider these facilities as suitable targets, and hence the probability of an attack needs to be reduced. The example clearly demonstrates how important it is to be precise regarding the background knowledge on which the subjective probabilities are based.

Terrorist attacks sometimes come as big surprises, such as the 11 September event in 2001. There were some signals and warnings that something might happen, but the event certainly came as a surprise to both professional security people and laymen. The same may be said of the killings in Norway on 22 July 2011, when a man placed a car-bomb outside the government office and massacred a number of people on the island of Utøya outside Oslo. We will discuss these events in more detail in the next section.

1.7 Black swans

In recent years the 'black swan' metaphor has become a very popular way of illustrating the idea of surprising events and outcomes. The black swan concept was first introduced by the Latin poet Juvenal, who wrote "rara avis in terris nigroque simillima cygno" (a rare bird upon earth, and exceedingly like a black swan), although according to Hammond (2009) that was imaginative irony. Juvenal's phrase was a common expression in sixteenth century London, used to describe something impossible. Up to that point in time, all observed swans in the Old World had been white. Taleb (2007, p. xvii) writes:

> Before the discovery of Australia people in the Old World were convinced that all swans were white, an unassailable belief as it seemed completely confirmed by empirical evidence.

However, in 1697 a Dutch expedition to Western Australia led by Willem de Vlamingh discovered black swans on the Swan River, and the concept of black swans developed to mean not only something extremely rare (a rarity) but also incorporating the idea that a perceived impossibility might later be disproven: a logical fallacy, meaning that if one does not know about something, it is therefore impossible. Taleb (2007) comments that in the nineteenth century the black swan logical fallacy was used by John Stuart Mill as a new term to identify falsification; he wrote: "No amount of observations of white swans can allow the inference that all swans are white, but the observation of a single black swan is sufficient to refute that conclusion." It became a classic example in elementary philosophy (Hammond 2009).

In 2007, Nassim Nicholas Taleb further defined and popularised the concept of black swan events in his book *The Black Swan* (Taleb 2007) (a second edition was issued in 2010 with a new section, which discusses various aspects of the probability concept, among other things). Taleb refers to a black swan as an event with three attributes. First, it is an outlier, falling outside the realm of regular expectations because nothing in the past can convincingly point to its possibility. Second, it carries an extreme impact. Third, in spite of its outlier status, human nature makes us concoct explanations for its occurrence after the fact, rendering it explainable and predictable.

Taleb's (2007, 2010) book has inspired many authors; for example, see the many references in Taleb (2013). However, some scholars are sceptical of Taleb's work. Professor Dennis Lindley, one of the strongest advocates of the Bayesian approach to probability, statistics and decision-making, has made his view very clear in a review of Taleb's book (Lindley 2008): Taleb talks nonsense. Lindley lampoons Taleb's distinction between the lands of Mediocristan and Extremistan, the former capturing the placid randomness of tosses of a coin, and the latter covering the dramatic randomness that provides unexpected and extreme outcomes. Lindley provides an example of a sequence of independent trials with a constant unknown chance of success – clearly an example of Mediocristan. Each trial is to be understood as a swan, and success is a white swan. Using simple probability calculus, Lindley shows that a black swan is almost certain to arise if you see a lot of swans, although the probability that the next swan observed is white is nearly one. Lindley cannot be misunderstood: "Sorry, Taleb, but the calculus of probability is adequate for all kinds of uncertainty and randomness."

In this book we will provide a thorough analysis of this issue: the concept of the black swan in relation to risk. What is the meaning of this term and metaphor in a professional/scientific setting? We question the extent to which the ideas of Taleb, and in particular the distinction between Mediocristan and Extremistan, can be given a proper scientific justification in view of existing risk theories and perspectives. Clearly, if Taleb has made some important points, Lindley cannot be right. More specifically, we discuss to what extent a black swan is:

I   A so-called unknown unknown (with extreme consequences). The United States Secretary of Defense, Donald Rumsfeld, made the term 'unknown unknowns' familiar to us all on 12 February 2002 at a press briefing. Addressing the absence of evidence linking the government of Iraq with the supply of weapons of mass destruction to terrorist groups, he said:

> There are known knowns; there are things we know we know. We also know there are known unknowns; that is to say we know there are some things we do not know. But there are also unknown unknowns – the ones we don't know we don't know.

An example of an unknown unknown is the occurrence of various new types of viruses throughout history.

However, the term was used long before Rumsfeld's popularisation. For example, it is mentioned by Furlong (1984), and in relation to climate change it has been commonly used for many years (e.g. Myers 1993).

II   A surprising extreme event relative to one's beliefs/knowledge.
III A surprising extreme event with a very low probability (extreme in the sense that the consequences are large/severe).

In Table 1.1 an attempt has been made to classify the various events discussed in previous sections according to these definitions of a black swan. The yes/no scores reflect that the events are black swans for some people and according to some interpretations, and not others. We will discuss the topic in detail in coming sections; for example, see Section 3.4.

Taleb (2007) provides a number of examples of black swans:

> Just imagine how little your understanding of the world on the eve of the events of 1914 would have helped you guess what was to happen next. (Don't cheat by using the explanations drilled into your cranium by your dull high school teacher.) How about the rise of Hitler and the subsequent war? How about the precipitous demise of the Soviet bloc? How about the rise of Islamic fundamentalism? How about the spread of the Internet? How about the market crash of 1987 (and the more unexpected recovery)? Fads, epidemics, fashion, ideas, the emergence of

art genres and schools. All follow these Black Swan dynamics. Literally, just about everything of significance around you might qualify.

(Taleb 2007, p. x)

Few of these events can be characterised as unknown unknowns, but they certainly were not easy to foresee. Taleb makes an interesting comment on the September 11 event:

Think of the terrorist attack of September 11, 2001: had the risk been reasonably *conceivable* on September 10, it would not have happened. If such a possibility were deemed worthy of attention, fighter planes would have circled the sky above the twin towers, airplanes would have had locked bulletproof doors, and the attack would not have taken place, period. Something else might have taken place. What? I don't know. Isn't it strange to see an event happening precisely because it was not supposed to happen? What kind of defense do we have against that? Whatever you come to know (that New York is an easy terrorist target, for instance) may become inconsequential if your enemy knows that you know it. It may be odd to realize that, in such a strategic game, what you know can be truly inconsequential.

(Taleb 2007, p. x)

These comments raise a number of questions. What does it mean that risk is reasonably conceivable? Can risk assessment identify black swans? And how shall we confront such events – i.e. how shall we manage the risks related to black swans? These are some of the main topics of the present book.

Table 1.1 Classification of some events as black swans in line with Taleb's (2007) definition and definitions I to III above

Interpretation of a black swan / Situations/events	A black swan as defined by Taleb	I – A black swan is an unknown unknown (with extreme consequences)	II – A black swan is a surprising extreme event relative to one's beliefs/ knowledge	III – A black swan is a surprising extreme event with a very low probability
Pickpocketing at Paris Gare de Lyon	No	No	No	No
Delivering a talk	No	No	No	No
The Macondo accident	Yes/No	No	Yes/No	Yes/No
The Fukushima Daiichi nuclear disaster	Yes/No	No	Yes/No	Yes/No
September 11	Yes/No	No	Yes/No	Yes/No
The attacks on 21 July 2011 in Oslo	Yes/No	No	Yes/No	Yes/No

It is obvious that the traditional probability-based approach to risk assessment and management does not provide the answer to meeting black swan types of events. Instead, we often refer to what we can call 'robust approaches', which are methods that are effective in meeting hazards/threats, surprises and the unforeseen. These approaches cover cautionary measures such as: designing for flexibility (meaning that it is possible to utilise a new situation and adapt to changes); implementing safety barriers; improving the performance of barriers by using redundancy, maintenance, testing etc.; and applying quality control/assurance. This term also covers concepts such as resilience engineering, which is concerned with finding ways to enhance the ability of organisations to be resilient, in the sense that they recognise, adapt to and absorb variations, changes, disturbances, disruptions and surprises (Hollnagel *et al.* 2006). We may also include the concept of antifragility (Taleb 2012). According to Taleb, the antonym of fragile is not robustness and resilience, but 'please mishandle' or 'please handle carelessly'. The illustration Taleb uses is sending a package full of glasses by post. The antifragile is seen as a blueprint for living in a 'black swan world', the key being to love randomness, variation and uncertainty to some degree, and thus also errors. Just as our bodies and minds need stressors to be in top shape and improve, so do other activities and systems.

In practice, for example in relation to industrial activities such as nuclear and oil and gas, the standard risk management approach represents a balance between probability-based risk management and robust methods. It is acknowledged that probabilistic approaches have limitations in managing risk, surprises and the unforeseen, and need to be supplemented by robust approaches. For societal safety and security contexts we see this duality even more strongly, for example in relation to terrorism risk. Here, probabilistic risk assessments are hardly used. The information provided by assigned attack probabilities is small in most cases, as discussed in the previous section.

This acknowledgement implies a need for a way of thinking about risk that sees beyond the probabilistic perspective. In engineering applications, risk is commonly conceptualised as expected values (probability multiplied with loss) or as the combination of probability and consequences (loss); see Section 2.2. However, such a perspective is too narrow to adequately capture the black swan type of risk, and in recent years we have seen many attempts to conceptualise risk to meet this broader view, as shown in Section 2.3. One example is the new definition of risk adopted by the ISO (2009a, b): "risk is the effect of uncertainties on objectives". A key aspect of this definition of risk is that uncertainty replaces probability. This may seem to be a rather minor change, but it has important implications. It will be thoroughly discussed in this book.

One may question why it is important to discuss the meaning of a black swan. Why should we try to agree on the definition of a black swan as a scientifically-based term in this field?

As a response to such questions, I would like to highlight that what is important is not really the term, but rather the related analysis and management. However, the concept of the black swan exists – the idea has gained a lot of attention and is a hot topic in many forums that discuss safety and risk. As a scientific and professional environment we cannot just ignore this. We need to provide perspectives and guidance on what this concept is saying. We need to place this concept in the frameworks that the risk field has developed over the years. This is exactly what we are doing in this book. Moreover, the risk field needs suitable concepts for reflecting this type of phenomena. The popularity of the black swan concept clearly demonstrates this, but I would also like to add that from a strictly professional point of view, there is a need for concepts that describe what a 'surprise' really means in this context. These concepts cannot and should not be limited by probability-based thinking and ideas, as the phenomena that we are trying to characterise extend beyond this paradigm. In the extended non-probability context we need to develop proper terminology – the present situation is rather chaotic – and in my view the black swan concept represents a useful contribution to this end. We can use statements such as those in definitions II and III above, but when we are communicating and discussing issues linked to surprising events, my experience is that it is very helpful to have at hand a term like 'black swan', which people can easily relate to. Using the black swan concept, I have noticed increased interest and enthusiasm for discussing risk issues.

In addition, I am convinced that studying the black swan concept provides new insights into the risk field regarding the links between risk, probability and uncertainties. The book is based on the conviction that there is a need to strengthen the foundation of the risk discipline by providing new insights into the relationship between surprising events, risk, probability and uncertainty. For this discipline, as for all other disciplines, it is essential that the conceptual basis is solid. However, this analysis is not only of theoretical and foundational interest. The main contribution of the work is not the definition of a black swan as such, but the structure developed to understand and analyse the related features of risk and uncertainties in a risk assessment and risk management context.

A simple demonstration of this is the example of Lindley (2008), as mentioned above. According to Lindley, there is no need for any uncertainty assessment that extends beyond the probabilistic. Taleb and many others, including the present author, reject this idea and seek to build a scientific platform for a more complete approach to risk and uncertainties. The present work can be seen as a contribution to this end. The analysis in the book is closely linked to the discussion of the role of risk assessment in risk management. Risk assessment can never fully capture black swans, but improvements can and should be made compared to the probabilistic approach that dominates the present quantitative risk assessment practice. To this end, it is considered essential to establish risk–uncertainty frameworks that are so broad that they also capture such events – such as, what was unknown at time t could be known by time s, and what is unknown to persons x could be known to persons y. The knowledge dimension needs to be highlighted much more strongly than is typically seen in risk assessment applications today.

In my view, Taleb's book represents an important contribution in this respect. It has been an aim of the present work to providing a stronger basis for his ideas and related work by relating them to key concepts such as risk and probability.

2   The risk concept

There is no widely agreed definition of the concept of risk. If we study the literature we find a number of different ways of understanding the risk concept. Some definitions are based on probability, chance or expected values, some on undesirable events or danger, and others on uncertainties. Some consider risk as subjective and epistemic, dependent on the available knowledge, whereas others grant risk an ontological status independent of the assessors. Section 2.2 of this chapter reviews the definition and meaning of the concept of risk in a professional context. The review includes a historical and developmental trend perspective, also covering recent years. It is questioned whether, and to what extent, it is possible to identify some underlying patterns in the way risk has been and is being understood today. First, in Section 2.1 we provide some remarks on the origin of the word 'risk' and how this concept is used in everyday language. Judgements about the match between the formal definitions and the phrases presented in Section 2.1 are provided to see the extent to which the formal definitions are in line with the typical daily risk language. Based on the review in Section 2.2, a recommended framework for understanding risk is presented in Section 2.3, where concepts such as hazard, risk source and vulnerability are also defined. Finally, Section 2.4 shows how this framework applies to the examples introduced in Chapter 1.

2.1 The origin and daily use of the risk term

The risk term has been used historically, and is also used today, in three different ways (*Oxford English Dictionary* 2014):

(1) (Exposure to) the possibility of loss, damage, injury, or other adverse or unwelcome circumstance; a chance or situation involving such a possibility.
(2) A hazardous journey, undertaking, or course of action; a venture.
(3) A person or thing regarded as likely to produce a good or bad outcome in a particular respect; a person or thing regarded as a threat or source of danger.

Here are some examples (*Oxford English Dictionary* 2014):

Category (1)
1699   They must run the risque of the falling of the price.
1718   He that runs the risque deserves the fair.
1808   Little he loves such risques, I know.
1880   Fire insurance as a business consists in undertaking a certain risk in return for a comparatively small sum, called the premium.
1889   Sugar is very difficult to ship; rum and tafia can be handled with less risk.
1908   The plan was rather a hazardous one, but she was prepared to take risks.
1934   Particular districts may be open to the risk of flooding arising from heavy rainfall or the thawing of snow.
2003   In Colorado, about 1.3 million people live in the red zone, areas at great risk of wildfire.

Category (2)
1679   Unto far distant Orbs, she takes her flight, and wanders, without Keeper, out of sight. Return, return, to thy imprison'd shrine; and shamefully repent, this risque of thine.
1692   An insolent despiser of Discipline, nurtur'd into Impudence by a long Risque of Licence and Rebellion.

Category (3)
1867   Married men are usually the most desirable risks.
1867   Gasworks are esteemed a fire risk of special hazard.
2001   Customers of other Citigroup companies who agreed to buy Travelers' insurance tended to be poor risks.
2002   The boy, who is considered a risk to himself and others, is currently being housed in a £3,500-a-week secure unit in Leeds.

The *Oxford English Dictionary* (2014) also provides some examples of the use of expressions like 'at risk', 'in risk', 'at a person's risk', 'at the risk of', and 'at risk to':

At risk, in risk
1796   The reputation of the whole fraternity would be at risk by irregularity on this occasion.
1937   Hundreds of Irishmen were glad to put their necks in risk of England's halter.
1994   Researchers worldwide have raced to identify more of the genes that put individuals at high risk of developing the disease, also known as juvenile diabetes.

At a person's risk
1852   If a part of the property insured was sold, it ceased thereby to be at the risk of the underwriters.
1922   The goods were carried at the owner's risk.
1970   He therefore should be at risk where the car causes damage.

At the risk of
1815   Ay, at the risk of all our own necks – we could do that without you.
1816   He ran down to the cellar at the risk of breaking his neck.
1874   You thought you could save him at the risk of your health.
2003   In science as in normal life, there are some clochards who, at the risk of being ridiculed, explore unpopular territories.

At risk to
1905   The bravery of eight men of the regiment who, at risk to their lives, snatched from the zone of fire a popular young officer.
1969   Increasing numbers of townsmen were engaged in forming politically oriented professional unions at considerable risk to the individuals involved.

These expressions are all linked to the first category of risk interpretations, (1) above. Let us consider an example to further illustrate the different meanings. We may imagine a scenario in which a person, let us call him Peter, is placed below a boulder that may or may not dislodge from a ledge and hit him; see Figure 2.1.

In this example we may state that:

(a) The person standing below the boulder (Peter) is exposed to (a) risk.
(b) There is a risk that the boulder will hit Peter.
(c) Peter takes (a) risk by placing himself under the boulder.
(d) Peter is in a risky situation.
(e) Peter risks getting hit by the boulder.
(f) Peter faces risk – the boulder may dislodge from the ledge and hit him.
(g) The risk that Peter faces (is exposed to) is small.

These examples are also illustrations of the first category of interpretations – category (1) above – as are the following sentences, addressing some current issues:

(A) There is risk associated with the operation of a nuclear power plant.
(B) Emergent risks have become an issue.
(C) The terrorism risk is high.

Figure 2.1 A person (Peter) standing below a boulder (based on Rosa 1998, Aven *et al.* 2011)

If we look at these examples we see the following similar features:

For an activity, different consequences are possible; one or more are negative (undesirable) and the main focus is on these – the consequences are not known. Risk is either the possibility/uncertainty/chance that the activity will have some undesirable consequences, or the activity (person, gasworks) itself, which is often also referred to as a risk source or a threat.

These current daily life interpretations of the risk concept match very well the original and historical use. According to Bernstein (1996), the word 'risk' derives from the early Italian *risicare*, which means 'to dare'. It was used by ancient sailors to warn the helmsman that rocks might be near. However, reading the analysis provided by the *Oxford English Dictionary* (2014) and Althaus (2005), it is clear that the origin of the word 'risk' is disputed; the etymology of the concept of risk is inconclusive. Here are some of the explanations provided by the *Oxford English Dictionary* (2014) (these are to a large extent in line with those mentioned by Althaus):

- French *risqué*: danger or inconvenience, predictable or otherwise (1578 in Middle French as a feminine noun, 1633 as a masculine noun; 1690 as a legal term)
- Italian *risco* (first half of the fourteenth century), variant of *rischio* (1292; thirteenth century as *reisego*), *risico* (1367), both in the sense of 'possibility of harm, an unpleasant consequence, etc.'

- Post-classical Latin *resicum, risicum* (both mid-twelfth century in Italian sources; also mid-twelfth century in a document from Constantinople), *risigum, resigum, resegum* (1227–8 in Occitanian sources), *rischium, rischum, riscum, risecum* (second half of the thirteenth century in Italian sources), all in commercial contexts in the sense of 'hazard, danger'
- Middle French (Walloon) *resicq, risicq*: possibility of damage to or loss of merchandise (second half of the fifteenth century)
- Old Occitan *rezegue*: possibility of damage to merchandise when transported by sea (1200; 1301 as *reseque*)
- Catalan *risc, reec*: danger, possibility of damage to merchandise when transported by sea (thirteenth century)
- Spanish *riesgo*: conflict, disagreement (1300), possibility of harm or unpleasant consequences (sixteenth century)
- Portuguese *risco* (fifteenth century)
- Dutch *risico* (1525) and *resicq* (1563), *risicque* (1602), both recorded in the Spanish Netherlands in the sense of 'possibility of damage to merchandise'
- German *Risiko* (1507)
- Post-classical Latin *rixicum* (1458 in an English source)
- Medieval Greek ῥίζικον (1160)
- Arabic *rizq* (in Maghribi Arabic also *rēzq*), which has a number of senses: 'provision, lot, portion allotted by God to each man', 'livelihood, sustenance', hence 'boon, blessing (given by God)', 'property, wealth', 'income, wages', and finally 'fortune, luck, destiny, chance'

According to the *Oxford English Dictionary* (2014), it is widely suggested that the post-classical Latin *resicum, risicum*, etc., in the sense of 'danger, hazard', originated from the post-classical Latin noun *resecum,* a possible derivative of the classical Latin *resecāre* meaning 'that which cuts' and hence 'rock, crag, reef' (compare Spanish *risco* in this sense from the thirteenth century), with allusion to the hazards of travel or transport by sea. The *Oxford English Dictionary* (2014) states that this argument fits with the maritime context of many early uses of the word in English and the Romance languages, but it involves a number of steps which are not supported by documentary evidence.

Again, following the *Oxford English Dictionary* (2014), this explanation, as well as one suggesting that the post-classical Latin *resicum, risicum*, etc. is derived from the specific senses of 'fortune, luck, destiny, chance' for the Arabic *rizq*, assume that the medieval Greek ῥίζικον was borrowed from the post-classical Latin *risicum,* but it is also possible that the borrowing went the other way: both words are first attested at about the same time. See the *Oxford English Dictionary* (2014) for some possible ways of explaining the origin of the Greek word (one is the Arabic *rizq*).

Althaus (2005) includes some interesting reflections concerning the variability in use of the word 'risk' across time, society, and region. Althaus refers to Bernstein (1996) and Gigerenzer *et al.* (1989), who to a large extent align the concepts of 'chance' and 'probability' with risk, and argue that the notion of fate was replaced with belief in the ability of humanity to master uncertainty using the tool of probability. Following this argument, any distinction *between* risk and uncertainty/chance today has been linguistically lost.

At the same time, however, risk is a very loose term in everyday parlance, and issues of calculable probability are not necessarily important to the colloquial use of risk (Lupton 1999). As will be seen from our later analysis, many risk perspectives make a sharp distinction between uncertainty/chance and risk.

Althaus (2005) concludes that the older entrepreneurial concept of risk as a venture has blurred since the beginning of the nineteenth century (Lupton 1999). In contemporary times the word 'risk' in everyday language has increasingly come to refer to something negative (Wharton 1992). These changes in the meanings and use of risk – resistance against fate, the merging of risk with uncertainty, and the contemporary emphasis on risk linked to undesirable consequences or outcomes – are associated with the emergence of modernity, beginning in the seventeenth century and gathering force in the eighteenth (Lupton 1999, Giddens 1999).

2.2 Development of a risk concept to be used in a professional context

This section reviews and discusses ways of defining and understanding the risk concept in a professional context. We cover concepts based on expected values, probabilities and uncertainties, and use several examples, including the boulder case, to illustrate ideas. The final part includes a discussion in which an attempt is made to categorise the various definitions of risk in a development scheme. It is argued that there has been a gradual change from rather narrow risk perspectives, based on probabilities and expected values, to broader non-probability-based definitions with a sharp distinction between risk as a concept and how this concept is measured. Many of the definitions studied are from the last thirty to forty years, but some definitions come from the first part of the twentieth century, and some from even earlier. The risk definitions of the last thirty to forty years have paralleled the development of the scientific field of risk analysis (see e.g. Thompson *et al.* 2005). A summary of key risk definitions with references is provided in Appendix B.

2.2.1 Expected values. Different types of probabilities

As we have seen, the origin of the term 'risk' is disputed in the literature. However, there seems to be broad agreement between researchers that De Moivre's 1711 definition is one of the first formal definitions of risk used in a risk analysis context (De Moivre 1711). De Moivre defines the risk of losing any sum to be the product of the sum adventured multiplied by the probability of the loss, i.e. risk is defined as the expected loss. This definition is still used in many contexts. Think of an insurance company that covers a huge number of assets, each worth 100000 and having a loss probability of 1/10000. Then the expected loss for the company equals 100000 × 1/10000 = 10, which can be seen as an informative risk index for the company when determining the premium.

However, the use of expected values is more problematic in other contexts. Let us try to use it in relation to the boulder case introduced in the previous section. Remember the statements (a) to (g) – for example, f and g:

(f)   Peter faces risk – the boulder may dislodge from the ledge and hit him.
(g)   The risk that Peter faces is small.

Loss in this case is injury or death caused by the boulder. To simplify, say that the only relevant loss is being killed. Then, expected loss can be replaced by the probability of this event occurring. Hence, we can rephrase the above boulder sentences:

(f)   Peter faces a probability of being killed – the boulder may dislodge from the ledge and hit him.
(g)   The probability of being killed that Peter faces is small.

Hence, to explain the risk concept in this case we need to understand what a probability means. Look at sentence (g) first. It makes sense. Intuitively, a low risk in this case corresponds to a low probability of being hit and killed by the boulder. This is true whether we interpret a probability as a judgement made by the assigner of the probability, or as a property of the situation under consideration. In the former case, the probability expresses the degree of belief of the assigner, a way of expressing his/her uncertainty about whether Peter will be killed or not. If a probability of (say) 10% is assigned, the assigner has the same degree of belief and the same level of uncertainty about Peter being killed as randomly drawing a specific ball out of an urn that contains ten balls; see Appendix A. We refer to these probabilities as subjective probabilities or knowledge-based probabilities, or just probabilities. A knowledge-based probability is always conditional on some background knowledge, which includes data, information, assumptions and beliefs. We write $P(A|K)$, where A is the event of interest and K is the background knowledge, to express the probability of A given K. When K is clear, we write $P(A)$ for short.

In the latter case it is presumed that a probability exists which characterises the situation – an objective property of the situation in the sense that if we could repeat it infinitely under similar conditions, this probability would be equal to the proportion of times that the person would be killed.

Nevertheless, the meaning of such a frequentist probability is not clear. To define it we have to construct a population of similar situations. This can be done in many different ways; there is no objective approach for making this mental construction. This is one possible interpretation (Aven *et al.* 2011): boulders were placed on similar ledges a long time ago and the frequentist probability is the proportion of these boulders that dislodge and kill the person after a time period corresponding to Peter being underneath the boulder. The boulders may dislodge at different times, as they were placed on the ledge in different ways. It is not difficult to produce other interpretations, for example reflecting variations due to different climate and weather conditions (scenarios). The frequentist probability represents the fraction of the conditions (scenarios) for which the boulder dislodges when Peter walks underneath it. The point being made is that the model supporting the construction of the frequentist probability is subject to judgements and is obviously not objective, independent of the assessor. At best it can be classified as inter-subjective within a category of experts. An alternative interpretation may be based on physical modelling of the phenomena. The boulder would dislodge if the vertical forces (load) exceed the resistance (strength), and the problem is changed to describe the variations in the loads and strengths. Probability models are introduced to describe these variations – Gaussian distributions, for example. However, in this case there will also be a problem in explaining what the variations reflect. Does the load distribution reflect variation due to different positions of the boulder, or climate and weather conditions, or both? Hence, the physical modelling does not alter the problem of understanding what the frequentist probabilities express.

Frequentist probabilities are dependent on probabilistic modelling, and the assumptions underpinning the model as well as the functional relationships within the model must be justified. In the boulder example the model is not easily defined, and in other cases the whole idea breaks down. Think, for example, of the frequentist probability of a terrorist attack. The situation is unique, and it is impossible to define a population of similar situations in a meaningful way (Aven and Renn 2009).

In practice it is common simply to presume the existence of the frequentist probability, tacitly assuming that it reflects some variation in some population for similar situations. Stating that the probability that Peter faces of being killed is small may be sufficiently precise for a layman setting, but not for a professional one. We need to know what 'small' refers to. The frequentist probability is an unknown quantity and needs to be estimated by using relevant data and expert judgements. Then we need to be clear on the type of variation we are thinking about.

We must conclude that a frequentist probability does not exist in general; it is a model concept to be justified in specific cases. Hence, it cannot provide the basis for a general understanding of the risk concept. However, the degree of belief interpretation is universally applicable. See Appendix A for some historical remarks concerning this interpretation.

The distinction between conceptualisations that see probability as an objective property of the world and conceptualisations that are based on judgements and knowledge from a person (i.e. which are epistemological) has been thoroughly discussed in the literature (see bibliographic notes). It is linked to the well-known phrases used by Immanuel Kant (1724–1804), "Das Ding an sich" and "Das Ding fur mich". Probability and risk can be viewed as both an "an sich" property of the world and a "fur mich" concept; see for example discussions in Vatn (1998).

Now a few words about sentence (f): Peter faces a probability of being killed – the boulder may dislodge from the ledge and hit him. This sentence fits the idea that probability is a property of the situation, but not so much the degree of belief interpretation. An analysis of the other boulder statements, (a) to (e), provides similar conclusions.

Now let us return to the definition of risk as the expected loss. We have clarified what probability means. The question is then: to what degree does the expected loss provide a meaningful and useful definition of the concept of risk?

The general response to this question is that the use of expected values in risk management can seriously misguide decision-makers in practice. Two probability distributions may have the same expected values, one with mass centred around its expected value, the other having high probabilities for severe outcomes. Clearly the risk management for these situations should be different. For the latter probability distribution, there are 'high' probabilities for 'extreme outcomes' and cautionary measures such as emergency preparedness are normally required to meet these if one of them should occur. For the former probability distribution, such measures are not needed to the same extent.

Consider the following simple example. You have to choose between two project alternatives, I and II. The first alternative can lead to three outcomes, −1, 0 or 1, with probabilities 0.25, 0.50 and 0.25, respectively. The second alternative may lead to outcomes −1000, 0 or 1000, with the same probabilities. Then the expected value, the centre of gravity of these distributions, is 0 for both alternatives. Hence, if risk is given by the expected value, the risk is the same for these two alternatives. Clearly this way of measuring risk could be misleading, as the spread of the distribution is not captured by the risk measure.

Now, say that we consider many such projects which are independent of each other, and our focus is on the average outcome from these. Then we can apply a well-known result in probability theory, the so-called law of large numbers, which states that the average is approximately equal to the expected value (here 0). This result can be used as an argument for focusing on expected values when considering a set (portfolio) of projects, as major companies and states do. However, this argument is not as strong as it may seem at first glance. What is required is that the probabilities are frequentist probabilities and known, i.e. the variation in the phenomena must be known with certainty. Uncertainties and surprises have to be ignored. Of course, we cannot do this for most real-life problems. It would mean removing the most important aspects of risk, as will be argued for in later sections.

We have to conclude that expected values (loss) cannot be adopted as a general definition of risk, but of course it can be used as a risk index or metric, and depending on the situation it may be more or less informative and useful. Some common risk metrics based on expected values are PLL (Potential Loss of Lives), expressing the expected number of fatalities in a year, and FAR (Fatal Accident Rate), expressing the expected number of fatalities per one hundred million exposed hours.

The loss in 'expected loss' may be transformed to a scale of severity, expressing the magnitude of the loss on a relevant scale. The utility scale, as defined in expected utility theory, represents such a scale (Bedford and Cooke 2001). However, it is not easy to assign such a utility function for each potential outcome. Returning to the project example above, we have to transform the outcomes −1000, −1, 0, 1, 1000 to a utility scale, reflecting the utility of the outcomes for the assigner. In the case of expected utility theory, a [0, 1] interval is used, with the utility of −1000 being equal to 0, and the utility of 1000 being equal to 1. Then a utility score is assigned for the others according to a special lottery procedure. Take the outcome 0. If the assigner is indifferent between getting a certain outcome of 0 and playing a game which gives outcome −1000 with probability 1-p and outcome 1000 with probability p, the utility value of 0 is p. An overall expected utility value can then be computed by multiplying the probability of the various outcomes with the corresponding utility numbers and summing all these products.

Furthermore, whose utility function should be used? For many situations there are a number of stakeholders, and they may make completely different judgements about the importance of (say) environmental damage. Some may also be reluctant to reveal their utility function, as this would reduce their flexibility in negotiations and political processes. Should it not be possible to define and describe the risk in cases where the decision-maker is not able or willing to define his/her utility function?

The utility perspective reflects the need for seeing beyond expected values in risk management by incorporating the decision-maker's *risk aversion* or *risk-seeking attitude* in the utility (loss) function used. Risk aversion means that the decision-maker's certainty equivalent is less than the expected value; the certainty equivalent is the amount of payoff (e.g. money or utility) that the decision-maker has to receive to be indifferent between that payoff and the actual 'gamble'. A risk-seeking attitude means that the decision-maker's certainty equivalent is higher than the expected value (Levy and Sarnat 1994). Only in the case where the decision-maker is risk neutral can expected values replace the information provided by the whole probability distributions.

This discussion leads us to the distinction between analysis (evidence) and values, an issue that has been discussed thoroughly in the literature; see for example Renn (2008), Rosa (1998) and Shrader-Frechette (1991). Renn concludes that it is highly advisable to maintain the classic distinction between evidence and values, and also to affirm that justifying claims for evidence versus values involves different routes of legitimisation and validation.

However, the distinction between probability/risk and utility is disputed. Many of the earliest probabilists (e.g. de Finetti 1974) and many decision analysts link probability with utility in some sense. From their point of view, probability is an indispensable instrument for reasoning and behaving under uncertainty (de Finetti 1974); see also Merkelsen (2011). Such a perspective on risk and decision-making may work for some people in guiding their personal choices in life, but in my view it is naïve for applications in industry and society. If you are a decision-maker and would like to be informed by an expert expressing his or her subjective assessment of uncertainties about an event A, you would not be happy if this assessment was influenced by the expert's utility function or preferences concerning the stakes involved. The assessment would then be disturbed by external factors. It would also be unrealistic and unfeasible to force decision-makers to assign their probabilities and utilities in complex problems.

2.2.2 Risk as the pair of probabilities and consequences

About two hundred years after De Moivre's expected loss concept, we find risk being defined by probability; for example, risk is the chance of damage or loss (Haynes 1895). It is not difficult to find situations where this definition makes sense: for example, consider the probability of getting a specific disease, such as cancer. In health care it is common to talk about cancer risk, understood as the probability of getting cancer. We also remember the boulder case example studied in Section 2.2.1, where risk is the probability of Peter being killed by the boulder hitting him.

Both frequentist and knowledge-based probabilities are used; see the discussion in the previous section. However, the probability-of-loss dimension alone cannot be used as a general definition of risk. In one case, the consequences of getting cancer may be quite small, but in other cases they may be life-threatening. Risk should also cover this aspect, and this leads to risk being understood as the two-dimensional combination of consequences (loss) and probability. Here are some common definitions following this idea:

- Risk is a measure of the probability and severity of adverse effects (Lowrance 1976).
- Risk is the combination of probability and the extent of consequences (Ale 2002).
- Risk is equal to the triplet (s_i, p_i, c_i), where s_i is the ith scenario, p_i is the probability of that scenario, and c_i is the consequence of the ith scenario, $i = 1, 2, ...N$ (Kaplan and Garrick 1981).

The Kaplan and Garrick definition has been dominant in the nuclear industry over the last thirty years. An example of a risk index in line with these definitions is risk matrices. A risk matrix shows combinations of consequence categories (for example loss of lives) and associated probabilities.

A growing number of researchers and analysts have found the probability-based approaches for understanding risk to be too narrow. The main arguments are:

(1) Assumptions can conceal important aspects of risk and uncertainties.
(2) The probabilities can be the same, but the knowledge they are built on may be strong or weak.
(3) They are too often based on historical data.
(4) Surprises occur relative to the probabilities.
(5) There is too much reliance on probability models and frequentist probabilities.
(6) Probability is just one of many tools which may be used to describe uncertainties; why give this particular tool such a special position?

A probability $P(A|K)$ is based on some knowledge K, which typically includes explicit or implicit assumptions. For example, in a risk assessment of an offshore installation, the probabilities can be calculated on the premise that no hot work is performed on the installation. However, in real life this assumption may of course be violated. Hence there is an element of risk/uncertainty not reflected by the probabilities. In practice, risk assessment cannot be performed without making a number of such assumptions.

To illustrate this we may think of a simple case where $P(A|X = x)$ is assigned, where the condition is that the unknown quantity X takes a value x. Clearly, the uncertainty about X is not captured by the probability value. In some cases we may of course incorporate this uncertainty by using the rule of total probability and compute the unconditional probability $P(A)$ using the formula $P(A) = \Sigma\ P(A|X = x)\ P(X = x)$, where we sum the values that X can take. However, in practice it is impossible not to make some assumptions on which the probabilistic analysis is founded.

The second point is a simple statement that we may have two situations, with identical probabilities, $P(A|K_1) = P(A|K_2)$, but in one case the strength of the knowledge is strong and in the other case it is weak. The probability number itself does not reveal this aspect, but should not the risk concept reflect this difference in some sense?

To further illustrate these first two issues, we consider the probability of an attack, $P(A|K)$, and write it as $P(A|\ F_1,\ F_2,\ ...)$, where the F terms are the relevant components of this knowledge. Some of these components represent assumptions; others represent more or less vague understanding of a phenomenon. Here are some examples of such Fs:

- It is assumed that an actor has the capacity to carry out an attack (F_1).
- It is assumed that a specific source reporting on the plans of this actor is 100% reliable (F_2).
- There are different sources giving different information about the actor's intention to carry out an attack. It is assumed that these sources are 100% reliable (F_3).

It is clear that these assumptions may conceal important aspects of uncertainty. The assumptions may turn out to be wrong. The information about possible attackers may increase greatly, but this does not necessarily mean that the P changes. The classical example is the case when we are completely ignorant and assign a probability of 0.5. Of course, this number could also be assigned when we have fairly strong knowledge about the event under consideration. The above analysis makes it clear that it is not enough just to report the probabilities to describe risk. The background knowledge is equally important.

This is especially important when there are many stakeholders. They may not be satisfied with a probability-based assessment expressing the subjective judgements of the analysis group; a broader risk description is sought.

The third point relates to the common use of probabilities; although the purpose of a probability is to express uncertainty or variation in relation to a future situation, it is often used simply as a function of the historical data observed. In this way important aspects of change and potential surprise are not reflected.

The surprise aspect is also important from a more general point of view. Based on some beliefs, one may assign a low probability of health problems occurring as a result of a new chemical, but these probabilities could produce poor predictions of the actual number of people that experience such problems.

Probability models based on frequentist probabilities constitute a pillar of the probabilistic approach. However, as discussed in the previous section, for many types of applications these models cannot be easily justified, with the consequence that the probability-based approach to risk and uncertainty becomes difficult to implement. A probability model presumes some sort of phenomenon stability; populations of similar units need to be constructed (see Appendix A.1.2). However, in a risk assessment context, the situations are often unique and the establishment of such models becomes difficult, as shown by the boulder example in Section 2.2.1.

By using probability as a main component of risk, we restrict risk to this measurement tool. In this view, probability is a tool introduced to describe/ measure the uncertainties. However, many other representations of uncertainty exist, including imprecise (interval) probability, and representations based on the theories of evidence (belief functions) and possibility (see Appendix A). In recent years such representations have received considerable attention among researchers and analysts; see the discussion in Appendix A.2. For these alternative approaches and theories, a probability-based risk definition cannot in general serve as a conceptual framework for defining and understanding risk. A broader risk perspective is required; probability has to be removed from the definition of risk, and the natural replacement is uncertainty.

2.2.3 Risk is uncertainty and objective uncertainty

The risk = uncertainty perspective is mainly linked to the economic field, and to investment analysis in particular. The qualitative term is 'uncertainty', and variance is a common tool for measuring uncertainties. The idea that risk equals uncertainty seems to be based on the assumption that the expected value is the point of reference and that it is known or fixed. The uncertainty is viewed in relation to known values – for example, a historical average value for similar investments. Risk captures the deviation and surprise dimension compared to this level.

Without such a reference level, the risk = uncertainty thesis does not work. Uncertainty seen in isolation from the consequences and the severity of the consequences cannot be used as a general definition of risk. Large uncertainties need attention only if the potential outcomes are large/severe in some respect. Look at the following example (Aven 2010a, p. 52): the activity considered can result in only two outcomes:

> 0 and 1, corresponding to 0 and 1 fatality, and the decision alternatives are A and B, having uncertainty (probability) distributions (0.5, 0.5), and (0.0001, 0.9999), respectively. Hence, for alternative A there is a higher degree of uncertainty than for alternative B, meaning that risk according to this definition is higher for alternative A than for B. However, considering both dimensions, both uncertainty and the consequences, we would, of course, judge alternative B to have the highest risk as the negative outcome 1 is nearly certain to occur.

> The expected values for the two alternatives are 0.5 and 0.9999, respectively, whereas the variances are 0.25 and 0.0001, respectively. The example is related to safety, but the conclusion is general and it also applies to investment problems. It does not make sense to say that there is negligible risk if we can predict with high confidence that the loss will be $10 billion. The uncertainties would be small, but it is misleading to use the term 'risk' for the uncertainties as such. The consequences dimension must also be taken into account.

A closely related understanding of risk is the perspective provided by Knight (1921). The idea is that we have 'risk' in the case where an objective probability distribution can be obtained (and 'uncertainty' otherwise). This thinking has strongly influenced the risk field, and in particular the economic risk area. However, it is difficult to find any good argument for using this terminology. It is not in line with the common risk language. Referring to risk only when we have objective distributions would mean excluding the risk concept from most situations of interest. If we adopt the subjective or Bayesian perspective on probability, Knight's definition of risk becomes empty. There are no objective probabilities. Given these observations, it is hard to understand why this definition is still being used. In my view we should abandon the Knight nomenclature once and for all. However, there is little evidence that my wish will soon be met. Economists and others continue to refer to this definition despite the strong arguments against it. They seem to see Knight's words as well established terminology (Aven 2010a, p. 75).

Of course, one may argue that the Knightian risk–uncertainty dichotomy is not so important as long as it does not affect the assessment and management of the totality of risk/uncertainty. The dichotomy is just a classification. This is true; however, the scientific risk fields cannot be based on a terminology that simply restricts itself to a set of more or less trivial situations and excludes the majority of cases.

2.2.4 Risk as an event or a consequence of an event

In the social sciences, two prevailing definitions of risk are:

> Risk is a situation or event where something of human value (including humans themselves) is at stake and where the outcome is uncertain .
> (Rosa 1998, 2003). (2.1)

> Risk is an uncertain consequence of an event or an activity with respect to something that humans value.
> (IRGC 2005)

These definitions are expressing that risk is a situation, an event or a consequence of an event. The first condition of risk is that the situations, events and consequences are subject to uncertainties. The second condition is that something of human value is at stake. Consider the activity of smoking. Humans have a stake in their health and well-being, and these values are threatened by events (such as the uncontrolled growth of cells) that could lead to lung cancer and death. The events and consequences (outcomes) are subject to uncertainty. Alternatively, we could consider lung cancer as the 'event' in the definition and reformulate this as: Humans have a stake in their health and well-being, and these values are threatened by lung cancer that could lead to weakened health or death. 'Lung cancer' is the risk, according to Rosa's definition (2.1). Risk is defined as an event (where...). 'Lung cancer' is the state of the world. The event is subject to uncertainties and threatens humans' health and well-being.

If, on the other hand, we refer to the 'situation' in definition (2.1), smoking can be considered as a risk. The outcome is uncertain and there are values at stake. For the IRGC definition, risk is the consequences of smoking.

Consider the boulder in Figure 2.1. Whether the precariously perched boulder will dislodge from the ledge is subject to uncertainty, and there are uncertainties about the consequences if the boulder does indeed dislodge. The boulder represents a threat to Peter who walks underneath it.

So what is risk in this example? According to Rosa's definition, the risk is the event that the boulder dislodges, which is characterised by a potential for killing or injuring Peter (something of human value is at stake), and this event and its consequences are subject to uncertainties. As with many of the threats that abound in modern society, Peter may be entirely unaware that the boulder is precariously lodged above him. Hence adopting such a risk perspective means that risk exists objectively. No one (with normal senses) would dispute that the event – the boulder dislodges – exists independently of the specific assessor.

Alternatively, if we instead focus on the 'situation' in definition (2.1), Peter walking underneath the boulder is a risk. For the IRGC definition, risk is the consequences for Peter if the boulder dislodges.

Now we would like to ask how to describe or measure this risk. Is the risk high? If risk is a situation, we quickly see that talking about high or low risks becomes impossible. The idea that risk is a situation leads to some conceptual difficulties. Rather, if we let risk be the situation, it seems more natural to refer to risk as something produced by an activity – smoking is the activity, walking underneath the boulder is the activity, etc. – and the risks are linked to these activities in some sense. Also, if risk refers to the event (for example, lung cancer, or the boulder dislodging and hitting Peter) or the consequences of the event, such conceptual difficulties arise. For example, we cannot conclude that the risk is high or low, or compare different options with respect to risk. Here are some examples of daily language phrases which do not make sense when interpretation (2.1) is adopted:

- In Colorado, about 1.3 million people live in the red zone, areas at great risk of wildfire.
- The goods were carried at the owner's risk.
- He ran down to the cellar at the risk of breaking his neck.
- The risk that Peter faces is small.
- The terrorism risk is high.

However, by focusing on the features that characterise the situation, namely uncertainties and what is at stake, we are able to define a general way of expressing the magnitude of the risk. We will do this in the next section.

2.2.5 Risk is the two-dimensional combination of consequences and uncertainties

For this risk perspective, risk is considered to be the two-dimensional combination of the consequences of an activity, C, and associated uncertainty (not knowing what the consequences will be), U. Here C is often seen in relation to some reference values (planned values, objectives etc.), and the focus is normally on negative, undesirable consequences. There is always at least one outcome that is considered as negative or undesirable. The consequences are with respect to something that humans value (including life, health, environment, economic assets). For short we refer to this risk as (C,U).

In general terms, risk is described by (C',Q,K), where C' are the specific consequences considered, Q is a measure of uncertainty (measure interpreted in a wide sense), and K is the background knowledge on which C' and Q are based. The most common method for measuring the uncertainty U is probability P, but other tools also exist, including imprecise (interval) probability and representations based on the theories of possibility and evidence; see Appendix A. The number of fatalities is an example of C'. Depending on what principles we adopt for representing C and the choice we make concerning Q, we obtain different ways of describing or measuring risk.

Think of the plant example in Section 1.3. Looking at the future operation (say next year), we are facing risk – for example, related to the loss of lives due to accidents. The actual consequences we call C, and they are unknown now. To describe the risk we have to specify the types of consequences that we will address – for example, the number of fatalities – and how we are going to express the uncertainties about these consequences. We have to determine which measure of Q to use (see Section 2.4.3). Note that the case of certainty (the consequences being a specific value, c) is a special (degenerate) case of uncertainty, characterised by $P(C = c) = 1$.

The consequences also cover events, such as a gas leakage or other hazardous events in a process plant. Let us return to the boulder case. Either the boulder dislodges from the ledge or it does not, and if the boulder dislodges the result could be that Peter is missed entirely, or hit by the boulder and injured or killed. All of these events are possible, but the occurrences of these events are not known, i.e. they are subject to uncertainties. These events and outcomes are states of the world, as was noted in relation to Rosa's definition in the previous section. They can be said to exist objectively, i.e. independent of the assessor – be that Peter or anyone else.

It does not seem natural to refer to the uncertainty component as a state of the world, but it does exist objectively in the sense that no one (with normal senses) would dispute that future events and consequences are unknown. 'Being unknown' is not dependent on your knowledge about these events; it simply reflects the fact that the future cannot be accurately foreseen. Hence, the uncertainty component also exists objectively in the sense of broad inter-subjectivity.

Consequently, according to this (C,U) definition, risk exists objectively (in the sense of broad inter-subjectivity). However, as soon as we view uncertainty as capturing knowledge beyond 'being unknown', the status of risk shifts to a subjective risk measurement (description). Hence risk as a concept exists objectively, and when risk is assessed it is dependent on the assessor – it becomes subjective. This is the essential nexus between the ontology and the epistemology of risk.

In this view, the (C,U) definition and Rosa's definition (2.1) can be seen to follow the same logic:

- Risk as a concept is based on events, consequences and uncertainties.
- Risk measurements are in line with the triplet (C', Q, K).

However, in the Rosa definition only the 'event' (the boulder dislodges) or the associated situation is considered to be a risk. The ontological status is the same, although the definitions are based on different ideas.

Hence there is an 'objective risk' of Peter getting hit by the boulder and being killed. The risk assessment may completely overlook this risk, but this does not alter the fact that Peter is facing this risk.

We see that the (C,U) way of understanding and describing risk allows for all types of uncertainty representations, and it could consequently serve as the basis of a unified perspective on uncertainties in a risk assessment context. In the risk descriptions various risk metrics can be used, for example based on expected values and probabilities, but the knowledge dimension (data, information, justified beliefs) and the strength of this knowledge need to be seen as integral parts. The risk descriptions may use modelling, such as probability models whenever they can be justified and are considered suitable for the analysis. If we use a probability model with a parameter p, this p is unknown and must be viewed as a component of C' in the general description of risk.

2.2.6 Discussion

Looking at the many categories of risk definitions, it is possible to extrapolate several development paths. Aven (2012b) has identified six, all starting from De Moivre's expected value (loss); see Figure 2.2. These paths are thought-constructed schemes reflecting some main types of the prevailing risk perspectives of today, with plausible development paths ending up with these current perspectives of today. Think about a modern person who argues for the use of the (C,P) perspective (D2 in Figure 2.2). If he/she had lived some x years back in time, what would his/her perspective have been? The scheme says that if the period x was between 30 and 110 years, the person is likely to have been ruled by the 'risk = probability of a loss' perspective – and so on. Alternatively, we may think of a set of six different categories of people, with members in each category being defined as 'similar' in some meaningful sense, even though they live in different time periods. One way of defining similarity is to refer to the same training or occupation, for example engineers or economists. It is stressed that this is a highly simplified scheme based on mental simulation with the purpose of illustrating some overall lines of potential development. The figure is not to be read as an accurate timetable for specific events.

The six development paths D1 to D6 (the parentheses include typical categories of advocators of these perspectives) are:

(D1) Since De Moivre, risk has been considered an expected value (loss) E. No changes in views (decision analysts and economists).
(D2) The concept of risk has developed from the E perspective, to P and finally to the (C,P) view which now prevails (engineers, health personnel).
(D3) The path is the same as D2, but (C,P) has been recently replaced by (C,U) (engineering-based scientists).
(D4) Early on, the risk concept changed from the E perspective to U and has not changed since then (business).
(D5) Early on, the risk concept changed from the E perspective to OU (Objective Uncertainty) and has not changed since then (economists).
(D6) The concept of risk developed from the E stand to cover both U and P, then U, Po and (C,P), and finally all of C, (C,U) and ISO (people with a pragmatic perspective), where Po refers to Possibility/potential, and C indicates that risk is an event or a consequence.

Within these perspectives we have interpretations based on both frequentist probabilities and subjective (judgemental, knowledge-based) probabilities.

The different paths should to a large extent be understandable in view of Sections 2.2.1 to 2.2.5. However, some remarks are appropriate.

Development path D6 is characterised by the acceptance of different perspectives on risk. Thinking in the first part of the twentieth century is characterised by both the P and U risk definitions, whereas the last part takes account of U, Po and (C,P). Recently the C, (C,U) and ISO definitions have been adopted. The underlying thinking for this development path is a pragmatic view that considers which risk perspective is the most suitable one.

An explanation is in order for the 'Risk = Potential/possibility of a loss' definition. Basically, this definition states that a loss may or may not happen (or a loss of different magnitude may occur), which is not far from saying that there is uncertainty about the loss (U), i.e. what the outcome will be – or perhaps even more accurately, that risk is (C,U) because the potential/possibility relates to different outcomes. It is also common to relate potential/possibility to probability – in some cases the word may be used with the same meaning, but perhaps a more reasonable interpretation is to relate these words to the idea of being 'thinkable' – an interpretation that seems

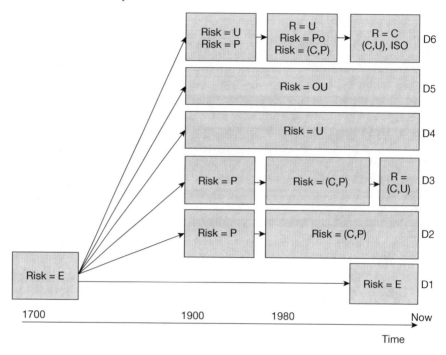

Figure 2.2 Six thought-constructed development paths for the risk concept.
E = Expected value (loss), P = Probability (of an undesirable event),
OU = Objective uncertainty, U = Uncertainty, C = Event/consequence,
Po = Potential/possibility (of a loss), ISO = ISO (2009a, b) definition of
risk (based on Aven 2012b)

also more in line with the use of the term 'possibility' in possibility theory to measure uncertainty (Dubois 2006).

It is possible to generate many other development paths. One obvious possibility would be to see risk as an event or a consequence (Risk = C) as a separate path, following E, P and (C,P). However, this path has not been shown in Figure 2.2, since the pragmatic view D6 seems to be the prevailing one among social scientists. Alternatively, we may define various sub-categories of D6, for example to reflect that Risk = C is to be seen as a ruling interpretation in recent years. We should bear this note in mind in the following analysis.

If we use the newest risk perspective from each of the six development paths, we end up with

E, (C,P), (C,U), U, OU and (C, (C,U) and ISO)

These represent the main categories of risk perspective which are in use today. These perspectives have developed in different ways; some go back over a century (three centuries for E), while others have been presented more recently. The different definitions can to some extent be traced back to different environments – economics, engineering and so on – but we see a recent tendency for more general, holistic concepts to be developed in order to capture requirements for assessing and managing decision problems, crossing traditional scientific disciplines and areas, and opening up new ways of describing/measuring uncertainties other than probability.

The discussion presented in Sections 2.2.1 to 2.2.5 above has led to the (C,U) perspective being the most appropriate type of risk definition in a general context. A short summary of the arguments follows.

The restriction of risk to objective uncertainties (OU) would mean an extremely narrow view of what risk is, and it is therefore rejected. The ISO definition is not sufficiently precise, but under certain interpretations it can be considered a special case of (C,U) or C. Seeing risk as uncertainty can be considered a special case of (C,U). Expected loss is rejected as a general risk concept as it does not adequately reflect the uncertainties. Expected (dis) utility is difficult to apply and it mixes uncertainty characterisations and value judgements, which is considered a disadvantage for practical risk management. Using the C concept would mean that we have to change the language of risk compared to everyday use. Therefore, what remain are the (C,P) and (C,U) categories of definitions. However, as argued in Section 2.2.2, the (C,P) perspective can be criticised for not taking into account the notion that probability is not a perfect tool, and the solution is to replace P by U to get (C,U). Risk as a concept should not be founded on one specific measurement tool. Following this argument, we obtain a sharp distinction between risk as a concept and risk descriptions (assessments), which could be based on models.

Risk definitions based on probability models (frequentist probabilities) are rejected, as such models make sense only in contexts of repeatability (refer to Section 2.2.2). For unique events, such models cannot be justified. However, they may play an important role in describing risk by supporting the assignments of subjective probabilities.

Despite the arguments for the use of (C,U) definitions, it is considered highly likely that all the perspectives – E, (C,P), (C,U), U, OU and (C, (C,U) and ISO) – will have a position in the risk field as founding pillars for a long time. For a specific context a particular risk perspective may work fine, and generalisations and more universal frameworks are not seen as attractive or necessary replacements. For actual risk assessment or risk management, it may be argued that how we theoretically define risk is not so important, as long as we are precise about what we describe when performing a risk assessment, there is consistency in terminology, and we are aware of and able to point to the limitations of the perspective adopted.

However, this line of argumentation fails because the risk perspective chosen strongly influences the way risk is analysed, and hence it may have serious implications for risk management and decision-making. Just being precise and consistent in terminology and aware of the boundaries and limitations of the tools that are used for the risk assessment is not enough. Some of the perspectives need to be wiped out, as they are simply misdirecting the decision-maker in many cases. The best example of this is the use of expected loss as a general concept of risk, but also purely probability-based perspectives should be included, since the uncertainties are not sufficiently revealed for these perspectives, as argued in previous sections. By starting from the overall, qualitative risk concept, we acknowledge that any tool we use needs to be treated as a tool. It always has limitations and these must be given due attention. By means of this distinction, we will be able to look for what is missing between the overall concept and the tool more easily. Without a proper framework clarifying the difference between the overall risk concept and how it is being measured, it is difficult to know what to look for and how to make improvements in these tools.

There is and should be a continuous discussion in scientific environments and application areas on how to best measure/describe risk. Most analysts would probably see the need for both quantitative methods and qualitative methods. Different situations (for example, with respect to the degree of uncertainties) call for different methods. In many cases a combination of quantitative and qualitative analyses would provide the best way of supporting decision-making.

As the above analysis demonstrates, the (C,U) risk perspectives (and also the majority of the others) are to a large extent in line with everyday language about risk and the etymology of the word 'risk'. The consequence dimension relates to something that humans value (health, the environment, assets etc.). The definition does not distinguish between positive and negative consequences (desirable and undesirable consequences), the point being that the activity results in some consequences (whatever these may be). However, in practice there are always some outcomes of C that are judged to be undesirable. The activity takes place, and although there is a possibility of undesirable consequences, the consequences could also be positive. When taking risk, we balance these concerns.

It would have been possible to restrict C to undesirable consequences. However, in doing this we need to determine what is undesirable, and for whom? An outcome could be positive for some stakeholders and negative for others. It may not be worth the effort and energy discussing whether an outcome is classified in the right category, and therefore most general risk definitions allow for both positive and negative consequences (Aven and Renn 2009).

In this book we are especially concerned about how the risk perspective supports the integration of concepts like unforeseen events, surprises and black swans. As we will see, the (C,U) perspective provides a good basis for this purpose.

The (C,U) perspective on risk is consistent with the view that decision-making under risk and uncertainties should be risk-informed, not risk-based. If one follows an approach where risk is identified with (for example) an expected value or a probability, risk-based thinking is not far away. On the other hand, in adopting the (C,U) perspective, where we distinguish clearly between the concept of risk and how it is described or measured, it is likely that a more humble attitude towards knowing the truth about risk will emerge. Risk-informed thinking becomes an integral part of the conceptual framework, which should increase the chance of a proper implementation in practice.

Of course, the risk-based/risk-informed issue also involves other matters, in particular the mindsets of risk analysts and risk managers concerning the use of risk assessments in the decision-making context. In practice we often see that risk managers want to be absolved of responsibility for having to make decisions – they apply a 'risk-based' approach in which their actions are dictated by the results of risk assessments. Then, if their actions turn out to be wrong or poor, they can claim absolution on the basis that: "We acted in line with the numbers. If the numbers were wrong, it is the analysts' fault." Risk analysts, for their part, often lack sufficient humility and fall into the trap of attempting to give answers with far greater certainty than can be justified. With the adoption of a perspective on risk like that recommended here, such a mindset may not disappear, but it would be more difficult to adopt, as it would lack justification in the way risk is understood.

Many researchers refer to risk but do not distinguish between risk per se and how risk is assessed and even managed. One illustrative example is provided by the German sociologist Ulrich Beck (1992). He states that: "Risk may be defined as a systematic way of dealing with hazards and insecurities introduced by modernization itself" (Beck 1992, p. 21). This represents a way of looking at risk which contradicts most other perspectives, and it is difficult to justify. To use the words of Campbell and Currie (2006, p. 151), in their investigation of Beck's work:

> It is hard to think of a less adequate definition: risk is not a way of dealing with things ... Beck's definition would make it impossible to ask: How are we responding to this risk?, as the response and the risk would be the same thing. Secondly, risk should not be so defined that it applies only to 'modernization', for there were of course risks before industrial society.

We also see that the concepts of risk and risk perception are mixed. The perception dimension includes personal feelings and affections (for example, dread) about possible events, the consequences of these events and their uncertainties and probabilities, and even judgements about risk acceptability. According to cultural theory and constructivism, *risk is the same as risk perception* (Jasanoff 1999, Douglas and Wildavsky 1982, Wynne 1992). Beck (1992, p. 55) states that: "Because risks are risks in knowledge, perceptions of risks and risk are not different things, but one and the same." This way of thinking also clashes with most professional and scientific risk perspectives, as they seek to distinguish between what is risk and what are feelings, emotions and value statements about risk (for example, what is acceptable risk).

2.3 Recommended conceptual risk framework

In general terms the recommended conceptual framework for understanding and describing risk is outlined in Section 2.2.5. Risk is defined by the pair (C,U), where C is the consequences of the activity considered, and U expresses the fact that these consequences are unknown. Here the consequences can be split into events A and consequences C, or risk sources RS, events A and consequences C, as shown in Figure 2.3. Hence risk can be defined by (A,C,U) or (RS,A,C,U) with corresponding risk descriptions (A',C',Q,K) or (RS', A',C',Q,K), with interpretations in line with (C',Q,K).

In more detail, the situation considered can be described as follows. We consider an activity, real or thought-constructed, for a specific period of time in the future, for example:

- the operation of a planned process plant;
- the society of a country;
- cars and pedestrians in a specific area;
- some bank operations; and
- the health condition of a person.

Some events A may occur, leading to some consequences C. Examples of such events in these examples are a fire, a terrorist attack, a pedestrian hit by a car, a shortfall in liquidity, and a disease (for example cancer), respectively. The consequences are with respect to something that humans value (health, the environment, assets, etc.). Normally the focus is on undesirable events and consequences, but the terminology also allows for desirable events and consequences. A severity scale is often introduced for characterising the magnitude of the consequences; for example, in relation to fatalities, the severity scale is simply the number of fatalities.

The events A can often be traced back to a source (a risk source), for example in relation to the five examples mentioned above:

- a maintenance activity in the process plant;
- suppression of democratic rights in the society;
- a slippery road;
- people unable to pay back their loans; and
- radiation in the building where this person works.

For each activity, we can define a related system – the process plant, the country (with its entire population), the pedestrian, the bank, and the person. The system may be exposed to the events, in the sense that the system is subject to these events; it can be affected by these events – there is a physical transformation of energy. The event may be initiated externally to the system or within the system. Only in the former case is it natural to talk about the system being exposed to the event; e.g. people are exposed to a terrorist attack, and the pedestrian is exposed to the car. The system may also be exposed to the risk source: the person is exposed to the radiation. However, if the risk source is a part of the system, as in the process plant example, it does not make sense to talk about exposure to the risk source.

We define a hazard as a risk source or an event that can lead to harm. A threat is understood to be a risk source or an associated event; most commonly used in relation to intentional acts affecting security.

To avoid events occurring and to reduce the consequences of the events if in fact they should occur, a number of barriers are introduced – for example, protection measures to reduce the effects of radiation. The occurrences of A and the values of C depend on the effectiveness of these barriers.

We are looking forward in time, and we do not know what events A will occur, and what the consequences C will be. There are uncertainties. There could also be uncertainties in relation to the risk sources: how intense will the radiation be? How will the suppression of democratic rights develop? What roads will be slippery, and to what extent? To what degree will people be unable to pay back their loans? How often will maintenance actions be carried out in the plant and what type will they be? The U in Figure 2.3 is simply expressing that we do not know with certainty what the future will bring. The real world is characterised by the risk sources RS, the events A, and the consequences C.

When considering the future, RS, A and C are unknown, and to provide a description (characterisation) of the future activity we need to specify a set of risk sources, events and consequences RS', A' and C', as well as a measure Q for measuring the uncertainties U. Here RS', A', C' and the measure Q are based on some background knowledge K, covering the assumptions, data and models used. Examples of RS', A' and C' are radiation dose rates at given times and places, the event that the person gets cancer, and the various outcomes if cancer occurs, respectively. Summarised, the future activity is described by (RS',Q,K), (A',Q,K) and (C',Q,K), or as an integrated description (RS',A',C',Q,K), noting that Q and K are used as generic symbols (they need not be the same in the different stages).

If we condition on the actual presence or occurrence of a risk source or hazard/threat, we talk about vulnerability. In line with the (A,C,U) definition of risk, this leads to the following definition of vulnerability:

> Vulnerability refers to the combination of consequences of the activity and associated uncertainties, given the occurrence of a specific risk source/event, RS/A.

Hence, if we speak about radiation risk for a person, there is uncertainty about the actual occurrence of and the amount of radiation, whereas if we refer to vulnerability, the exposure of radiation should be given/known. In the terrorist attack example, the societal risk captures the uncertainties related to whether or not an attack will occur, whereas for vulnerability this event has occurred and we are concerned about the consequences. Thus the vulnerability concept is risk conditional on the occurrence of a specific event A or a specific RS. When using the vulnerability term it is essential to refer to the relevant RS or A.

Following the adopted risk perspective, does it make sense to talk about *risk exposure* (as we often see in practice; see examples in Aven 2012c)? No, it does not. Think of the terrorist example. Can we say that the people are exposed to the terrorist risk, i.e. the attack and its consequences, with its uncertainties (A and C) unknown? Regarding the event, for example the attack, we can speak about exposure, but not consequences (for example, the radiation consequences), as these are not external to the system. We may incorporate the uncertainty component in relation to the event (the attack) in the exposure term, since it simply refers to the fact that we do not know whether or not this event will occur.

Next we need to consider ways of describing or measuring the vulnerability concept. The vulnerability description is similar to the risk description but conditional on the risk source RS or the event A (it is known that the system will be exposed to RS or A). Hence it takes the form $(C',Q, K \mid RS/A)$. Note that we are conditional on RS and A, presuming that they are known. Alternatively we could have written RS' and A'.

In the radiation example, the risk description captures the radiation doses (RS') with uncertainties assessed, and the effect that these have on the person – in other words the assessed system vulnerabilities given these doses.

If a system is judged to have high vulnerability, it is said to be vulnerable. The word 'robust' is often used for the case where the vulnerability is judged to be low.

Resilience is closely related to the concept of vulnerability and robustness. It is commonly referred to as the ability of a system to adjust its functioning prior to or following changes and disturbances, so that it can sustain operations even after a major mishap or in the presence of continuous stress. A main objective is to make the systems resilient so that they can withstand or even tolerate surprises (Renn 2008): "In contrast to robustness, where potential threats are known in advance and the absorbing system needs to be prepared to face these threats, *resilience* is a protective strategy against unknown or highly uncertain hazards"; "Instruments for resilience include the strengthening of the immune system; diversification of the means for approaching identical or similar ends; reduction of overall catastrophic potential or vulnerability; design of systems with flexible response options; and the improvement of conditions for emergency management and system adaptation."

To reflect and support this idea, the following definition of the resilience concept has been defined in line with the (C,U) risk concept:

Resilience = (C,U | any A/RS, including new types of A/RS), (2.2)

That is to say, resilience can be interpreted as the two-dimensional combination of consequences (C) and associated uncertainties (U) given the occurrence of any type of RS or A. An associated general resilience description can be formulated analogous to the vulnerability concept.

We are concerned about the performance of the system not only in the case of a specific RS or A, but for other risk sources and events as well. How will the system behave in the case of an event having some surprising features? Will the system be able to withstand or tolerate such events? To answer these questions we not only need to look at the states of the system and discuss how the system behaves under different types of stress; we must also address uncertainties and probability. We cannot design for all types of scenarios, and without incorporating the uncertainty and probability dimensions, judgements about resilience would not provide much guidance on this issue. The type of risk sources and events that we should include in the assessment (to be covered by the resilience concept shown in equation [2.2]) will always be an issue for discussion, because the forms of the surprises could be many. The key point here is simply that the issue is raised.

For the vulnerability concept, we also need to define some boundaries regarding which A events should be included in the definition. One may question whether we really need both of these concepts, vulnerability and resilience. Aven (2011c) discusses whether too many definitions could be a hindrance to professional practice and/or intellectual discourse. Obviously we should avoid having too many concepts and definitions, but which concepts and definitions are most appropriate? This is an important research question. The above nomenclature is motivated by the belief that it contributes to clarifications and a logically defined structure for risk, vulnerability and resilience. However, it is acknowledged that simplifications could be made. For example, we could completely remove the concept of resilience. 'Vulnerability' would then be the term used for reflecting 'risk' conditional on the occurrence of one or a set of events, A. A distinction is not made between whether the events are known or unknown. The point is that we always have to define the set of events on which the 'risk' is conditional when talking about vulnerability. A number of indices can be defined to measure vulnerability, but we do not need names for all types of such indices. Resilience is then captured by the concept of vulnerability/ robustness.

Models are used to analyse the consequences and the links to the risk sources and events. Of special interest are probability models. We will return to this issue in Chapter 3. The examples in the following sections, particularly Section 2.4.3, provide further details of this framework.

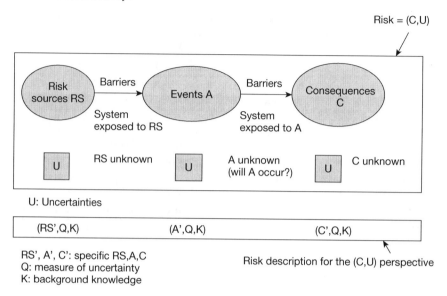

Figure 2.3 Main features of a conceptual framework for linking risk, risk sources and events in line with the (C,U) perspective (based on Aven 2012c)

2.4 Examples

In this section we provide some comments concerning risk and the related concepts discussed in Section 2.3, in terms of the examples introduced in Chapter 1.

2.4.1 Pickpocketing on the Paris Metro

When entering the Metro, I was facing risk. The consequences of taking the Metro could be that nothing special happens and I reach the airport as planned, or I could be robbed and lose my wallet. Of course, other events could also occur, but let us focus on the robbery. At the time when I enter the Metro, I do not know whether or not I will be robbed; the outcome is not known. There are uncertainties. There is risk. This risk exists even though I am not thinking about it.

The robbery represents the event A in this case. Boarding the crowded train is a risk source, and so is the robber. I did not assess the risk in this case, as we remember from Section 1.1, but let us assume that I did. How could I have done this? The first step would be to identify potential undesirable scenarios and events that could occur; a robbery on boarding the crowded train would be high on the list. Then a judgement of probability would be carried out, and the conclusion would be that this event represents quite a high risk unless measures are implemented – for example, a high level of awareness and avoiding entering the train if there is a large number of people in front of the doors.

To conclude, no quantification of the risk is necessary. Performing such a simple analysis would probably have been sufficient in this case to avoid the incident. I would have been aware of the threats and reduced the risk. Using the terminology in Section 1.2, I would have focused on failures, signals and warnings and been sensitive to operations; the result with high probability would have been a trip to the airport with no drama.

My assessment of risk has the purpose of supporting my decisions concerning my trip: when to take the train, and what risk-reducing measures to implement. Given my knowledge K, I find the probability of a robbery, $P(A|K)$ to be acceptable. This knowledge covers experience from many earlier uses of the Paris Metro, as well as general information about the safety/security of travelling on the subway, for example as reported through the media. If detailed statistics concerning robberies on the Metro were available, this information would be a part of the K as well. The fact that I consider the risk to be acceptable means that I judge the robbery probability $P(A|K)$ to be small. My decision-making does not require that I assign a specific number to this probability. It is not necessary; it would just be a numbers game, with no importance for the decisions to be made. Clearly, if numbers were to be assigned, intervals would be preferable – for example, expressing that the probability of a robbery without the implementation of risk-reducing measures would be between 0.01 and 0.1, while the probability with the measures implemented would be less than 0.01. I am not willing to express my degree of belief more precisely than this. To assign these intervals I make some assumptions, for example that I am not receiving and reading an email that disturbs me and renders me less focused on boarding the train. In line with the (C,U) risk perspective, the background knowledge (including assumptions) constitutes an important part of the risk description. We see that the introduction of quantified expressions of risk is rather complicated, and for a situation like this, a daily-life occurrence, it would not be needed for the decision-making.

However, if we view this decision-making from the perspective of the authorities, such numbers may be useful. The authorities may perform a risk assessment to support their judgements about risk acceptability and the possible need for measures to reduce risk. Using some historical data we can imagine that a risk matrix, like the one shown in Figure 2.4, might be presented. It shows, for example, that the probability of a robbery which only relates to economic loss is between $1 \cdot 10^{-5}$ and $1 \cdot 10^{-3}$ for a person who travels on the Metro in this area. The period considered is rush hour, comparable to when I was entering the train. These numbers are considered relatively high and the authorities decide that further measures are needed to reduce the risk, for example more use of guards in the Metro area and in particular trying to avoid chaos in the stage when travellers are boarding trains. This assessment also builds on background knowledge and important assumptions. As discussed in Section 1.1, there are a number of ways to define the relevant categories of data to be used for this assessment. Of

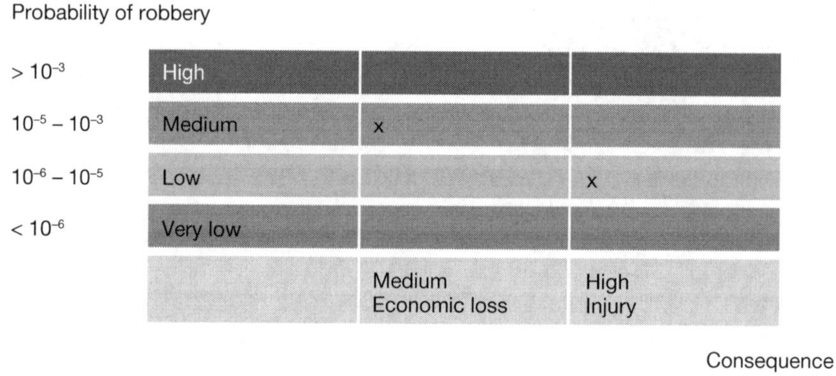

Probability of robbery

> 10^{-3}	High		
$10^{-5} - 10^{-3}$	Medium	x	
$10^{-6} - 10^{-5}$	Low		x
< 10^{-6}	Very low		
		Medium Economic loss	High Injury

Consequence

Figure 2.4 Risk matrix for the robbery example

course, the selection criteria need to be reported along with the risk matrix. The use of the interval instead of a precise number makes the assignment less sensitive to the exact selection criteria that are used. Overall, the knowledge that supports the quantitative analysis is judged to be strong in this case.

2.4.2 Delivering a talk to an audience of scientists

In Section 1.2 the concept of risk was explained in line with the recommended framework of Section 2.3: the talk can have different outcomes, and before it takes place we do not know which one will occur. This is risk. John has a focus on some undesirable scenarios, including situations that make him feel incompetent, but he also thinks about the positive outcomes: that the talk becomes a success. As in the pickpocketing case, the risk is not quantified. Of importance are the measures taken to meet the risk and achieve a good performance.

In planning for the talk, a number of hazards are identified (see Section 1.2). An example of a risk source is the presence of people who have presented work conflicting with John's. This risk source, as well as events such as 'some people leave the room', could easily lead to high vulnerability as John loses confidence. However, such events and conditions could also have the opposite effect: that John is motivated and steps up to the plate to meet these dangers.

2.4.3 Expressing and communicating risk – an offshore installation

We consider an activity, here the operation of an oil and gas installation offshore. The activity is real or thought-constructed, and is considered for a period of time from d_0 to d_2, where the main focus is on the future interval D from d_1 to d_2; see Figure 2.5. The point in time s refers to 'now' and

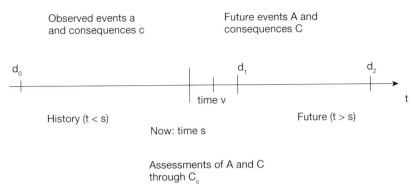

Figure 2.5 A schematic illustration of some of the fundamental components of the risk concept in relation to the time dimension. Here C_s refers to a set of quantities that is introduced to characterise the events A and consequences C in the period of interest, i.e. the interval D from d_1 to d_2 (based on Aven and Krohn 2014)

indicates when the activity is to be assessed or managed, as well as defining which part of the analysis can be regarded as history and which part as the future. If d_1 equals s, attention is on the future interval from now to d_2.

We may focus on the operation of the installation over its entire production period, or we may only be interested in the execution of a specific drilling operation at a specific period in time. Before the activity, at time s, we need a concept of risk, expressing in some way what could happen in the interval D – of events and consequences not as intended for this activity. A fire and explosion event may occur on the installation, and the drilling operation could lead to a blowout.

At time s, the activity performance in the future, with its events and consequences, is not known. There are uncertainties and risk. Hence, we need concepts that can help us to measure or describe these uncertainties, and the concept of probability enters the scenario. The uncertainties are linked to the knowledge of the assessor and are influenced by data and information gathered. If the assessor has moved forward to time v in Figure 2.5, he/she has updated knowledge and this would affect the risk assessments. At time v, signals and warnings of an accident may become available, and the challenge is to incorporate these into the risk concept in a way that makes the assessments informative and supportive of the decision-making. Risk can be described as outlined in Section 1.3 and further explained in Section 4.2.

The installation could be vulnerable to various hazards and threats – for example, if an explosion occurs. The vulnerability is described by the consequences that this explosion will have on the installation, with associated probabilities and uncertainties. Two examples of risk sources are the presence of hydrocarbons under high pressure and the many ships passing the installations.

Although we have referred to explosions and fires on an offshore installation, the above description is general; it explains the risk concept and its measurement, and highlights its link to the time dimension.

2.4.4 The Macondo accident

This example can be viewed as a special case of the one considered in the previous section. Well operations can lead to various consequences, for example a blowout and loss of lives, but we do not know in advance what the outcomes will be; there is risk. This risk can be described/measured in different ways; see Section 1.3 and Section 4.2.

2.4.5 The Fukushima accident

For this example, please refer to Section 2.4.3 and Figure 2.5. To understand the risk concept for this case, we have to consider the activity: the operation of the nuclear plant for a specific period of time. The consequences of the activity can be as planned, or an undesirable event may occur (such as a tsunami), leading to various losses. At time s we do not know if and when such events will occur. There are uncertainties; we are facing risk. The risk is typically described by probabilities and expected values; see Section 4.2 for further details.

2.4.6 Terrorism risk

We consider the lives of the inhabitants of a specific country. A terrorist attack may occur, leading to many injuries and fatalities. Today we do not know what the consequences will be, hence there is risk. Again we can refer to Figure 2.5. Clearly, people in that country do not only focus on the avoidance of terrorist attacks. They seek a 'good life', and there are many features of this life that could be negatively affected by measures aimed at reducing the likelihood of terrorist attacks. For example, think about the access to buildings and other places, which can be made very difficult in the presence of security measures. Hence the positive consequences of the activity cannot be seen in isolation from the negative consequences when managing the risk. Measuring terrorism risk is challenging, as indicated in Section 1.6.

2.4.7 Black swans

If risk is defined by (A,C,U), any black swan type of event would be covered by risk, as A and C are the actual events and consequences of the activity. If a new type of event occurs, it is included in A. However, it may not be included in A', the events on the risk assessment list. Provided the consequences of the event are extreme, a black swan type of event occurs if the real A is not covered by A'.

In Section 1.7 we also indicated that events with negligible probability could be seen as black swans. The events are known but are considered so unlikely that they are ignored – they are not believed to occur, and (additional) cautionary measures are not implemented. An example is the event that an underwater volcanic eruption occurs in the Atlantic Ocean leading to a tsunami affecting Norway. The events appear initially on the A' list, but they are then removed due to the fact that the probability is judged to be negligible; its occurrence will come as a surprise. The tsunami that destroyed the Fukushima Daiichi nuclear plant was similarly removed from the A' list due to its probability being judged as negligible.

It does not come as a surprise that a major accident, such as the Macondo event, occurs on an offshore installation if the perspective taken encompasses all offshore activities. A risk assessment of the whole industry would predict that several major events will occur in the next ten years. However, if we consider one specific installation, the particular events leading to a disaster may come as a surprise, a black swan for those that are involved in the management and operation of that particular installation. Remember all the events that went wrong in relation to the Macondo disaster (see Section 1.4). The probability that such an extreme event will occur is considered negligible. However, a number of robustness and resilience measures are implemented in preparation for extreme events, although the specific scenario is nearly impossible to foresee. The actual event and its consequences are specified by A and C, but A' and C' may capture these to a varying degree, because they are either not identified by the risk analysts or are ignored due to low probability.

Bibliographic Notes

This chapter is to a large extent based on Aven (2012b), which contains a review and discussion of key categories of risk, including the development scheme shown in Figure 2.2. The analysis of the ontological status of the risk concept is mainly taken from Aven *et al.* (2011). Figure 2.3 and the related discussion are based on Aven (2012c), while Figure 2.5, with its link between risk and time, is taken from Aven and Krohn (2014).

In their 1981 paper, Kaplan and Garrick (1981) refer to risk as qualitatively defined by "uncertainties + damage", which can be seen as a version of the (C,U) type of definition. However, these authors did not develop a theory for this perspective like the one presented here, as shown above by the (C,U) – (C'Q,K) approach (this theory is built on early work by, for example, Aven [2000] and Aven and Kristensen [2005]). For some further reflections on the ontological status of the concept of risk, see Rosa (1998), Aven *et al.* (2011) and Solberg and Njå (2012).

3 Variation, knowledge, uncertainty, surprises and black swans

This chapter provides a more detailed analysis of variation, knowledge, uncertainty, surprises and black swans in a risk assessment and management context. First, in Section 3.1 we look at variation and its link to probability modelling. A subsection is included on the distinction between 'common-cause variation' and 'special-cause variation', which are key concepts used to control and improve different types of processes in quality management. We study the meaning of these concepts, with a special focus on the latter notion: how is the special-cause concept linked to ideas and concepts used in risk assessment and management to reflect unforeseen and surprising events? In the quality discourse it is common to refer to two possible mistakes when confronting variation: (i) to react to an outcome as if it were the result of a special cause, when actually it resulted from common causes of variation; and (ii) to treat an outcome as if it were the result of common causes of variation, when actually it was the result of a special cause. However, at the point of decision-making, it is difficult or impossible to know what the 'true' state is. It is also appropriate to ask whether such a true state does in fact exist. In Section 3.1.1 we discuss these issues, the main aim being to improve our understanding of the variation concept and its link to risk and surprises.

The next section, Section 3.2, examines the knowledge concept in relation to risk. Common understandings of knowledge are reviewed and discussed. The classical definition refers to three criteria for a statement to be considered knowledge: it must be justified, true, and believed. The perspective adopted in this book is that knowledge is justified beliefs; any reference to a true statement is avoided. But how should we determine what is 'justified'? This is a topic that has been thoroughly discussed in the literature. The view taken here is that it is a result of a reliable process, and in this section we discuss what such a process means in relation to three cases: the assignment of a knowledge-based probability, quantified risk assessments, and the science of risk analysis. When it comes to quantified risk assessments, we discuss how the risk concept is able to reflect the knowledge dimension, more specifically the available data (D), information (I), knowledge (K) and wisdom (W), i.e. the various elements of the so-called DIKW hierarchy. A structure (conceptual framework) is described and discussed for linking risk and the DIKW elements. The structure is based on the following main ideas: Data = the input to the risk assessment; Information = the risk description; Knowledge (for the decision-maker) = understanding the risk description; Knowledge (for analysts) = understanding how to do the risk assessment as well as understanding the risk description; Wisdom (for the decision-maker) = the ability to use the results of the analysis in the right way; and Wisdom (for analysts) = the ability to present the results of the analysis in the right way.

Science is seen as a means to produce knowledge, but what does this mean in relation to risk analysis and its use? A main category of risk studies is about concepts, theories, frameworks, approaches, principles, methods and models to understand, assess, characterise, communicate and (in a wide sense) manage risk, in general and for specific applications. But are these studies scientific? These are some of the issues we discuss in this section. The aim is to provide a deeper understanding of the concept of knowledge in a risk context.

Section 3.3 studies the uncertainty concept in a risk assessment and management context. We distinguish between uncertainties about an unknown quantity, uncertainties regarding what the consequences of an activity will be, and uncertainty related to a phenomenon, for example in relation to cause–effect relationships. The challenge is to conceptualise uncertainty and then to measure it. The basic thesis is that uncertainty about a quantity or what the consequences of an activity will be constitutes not knowing the value of this quantity and not knowing what the consequences will be, and that measurement of this uncertainty leads to concepts like probability, interval probability and possibility, as well as characterisations of the knowledge that this measure is based on. For the uncertainty related to a phenomenon, various supporting concepts are studied, such as scientific uncertainties and lack of predictability. The concept of model uncertainty is also addressed. Finally, this section provides some reflections on the issue of "who is uncertain; is it the risk analysts or the experts"?

Section 3.4 provides some further reflections on the meaning of surprises and black swans in a risk context, following up the analysis in Sections 1.7 and 2.4.7, including Dennis Lindley's critique against the use of the black swan concept. Different perspectives on the understandings of a black swan are reviewed and discussed, taking into account both the time dimension of risk as well as the observer. We also look into other classifications, for example the ones based on unknown unknowns, unknown knowns, etc.

In the final Section 3.5 some overall conclusions are drawn in a discussion of the key points to be highlighted as input to the coming analysis on risk assessment and management in Chapters 4 and 5.

3.1 Variation

Figure 3.1 shows an example of leakage data for an offshore installation over a period of twelve months. The figure shows the variation in the number

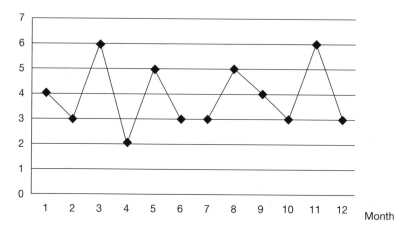

Number of leakages

Figure 3.1 Historical records of gas leakages on an offshore installation over a
period of twelve months (based on Aven 2014a)

of leakages for these months. In the first month we have observed 4 leakages,
then 3, and so on. On average, the number of leakages is 47/12 = 3.92, i.e.
approximately 4. The range is between 2 and 6, with a frequency of 1, 5, 1,
2 and 2 for the leakage numbers 2, 3, ..., 6 respectively.

The above analysis is an elementary descriptive study of observed data. It
provides a basis for more advanced statistical analysis with the purpose of
identifying trends and making predictions for the future. The common tool
for performing such an analysis is the use of a probability model, which is a
model of the variation of the quantity of interest, here the number of
leakages. A model is a representation of the phenomenon studied, here the
number of leakages, and it is a *probability* model because it is based on
probabilities. However, here we refer to not just any type of probability; it
is essential that the probabilities are frequentist probabilities (chances) as
defined in Section 2.2 and Appendix A.1.2. To keep things simple, let us first
consider a more simple case: a game of throwing a die.

Clearly if the die is a standard/ideal one, the variation will be given by a
chance/frequentist probability distribution (model) p_i, i = 1, 2, ..., 6, where
p_i = 1/6 is the fraction of times the die shows i when being thrown over and
over again a large number of times (in theory, an infinite number of times,
assuming the die does not wear out over time). Based on this probability
model, we will predict that in a number of throws, 1/6 will show a specific
number, for example 2.

Now suppose we have a die which is not necessarily fair. Then we can
establish a probability model expressing that the distribution of outcomes is
given by $(p_1, p_2, ..., p_6)$, where p_i is the chance of outcome i, interpreted as
the fraction of outcomes resulting in outcome i, i = 1, 2, ..., 6. This model

has unknown parameters p_j, and by statistical analysis we estimate these parameters and express uncertainties.

Two types of approaches are used for this purpose: the traditional frequentist approach and the Bayesian approach. The former is based on the use of frequentist probabilities, point estimations, confidence interval estimation and hypothesis testing, as outlined in any text book on statistics (see e.g. Anderson *et al.* 1994). On the other hand, the Bayesian approach is based on subjective (knowledge-based) probabilities expressing the analysts' judgements about the value of the parameters introduced. The idea of a Bayesian analysis is to first establish probability models that adequately represent the variation of the phenomena (in a risk assessment context this variation is sometimes referred to as aleatory or stochastic uncertainties; see e.g. Apostolakis 1990). The subjective probabilities are used to reflect epistemic uncertainties, i.e. incomplete knowledge or lack of knowledge about the values of the parameters of the models. At the initial stage of the analysis, these probabilities are referred to as prior subjective probabilities. When new data and information become available, Bayes' theorem is used to update the description of the epistemic uncertainties in terms of the posterior distributions. Finally, the so-called predictive distributions of the quantities of interest and the observables (for example, the number of system failures) are produced by applying the law of total probability. The predictive distributions are epistemic, but they also reflect the aleatory uncertainties. See any book on Bayesian analysis, for example Bolstad (2007) and Bernardo and Smith (1994), for further details.

Let us return to the leakage example. The data in Figure 3.1 indicate a process that is stable, with an average close to 4.0 events per month. To model the variation, the natural choice would be to use a Poisson process with a constant occurrence rate λ, estimated to be equal to about 4.0. Hence predictions can be derived for the number of events that will occur in the coming months. Following standard statistical theory, tests can be introduced to reveal when the process is out of control, for example where there is a significant deterioration of the process. Formally the idea is to define a null hypothesis, here (say) $\lambda = 4.0$, and consider the process to be out of control if $\lambda > 4.0$ or $\lambda < 4.0$. Observing the number of events X in a month, it is concluded that the process is out of control if $X \geq c_1$ or $X \leq c_2$ where c_1 and c_2 are chosen so that the test has a specific significance level (Oakland 2003, Bersimis *et al.* 2007). If a Bayesian analysis is carried out, a prior distribution is first assigned for λ, for example a gamma distribution. When new information is obtained, this distribution is updated using Bayes' formula to produce the posterior distribution. From this distribution we can calculate the subjective probability that, for example, λ is greater than 4.0 given the data observed. If this probability is small, the hypothesis that $\lambda = 4.0$ or larger is rejected.

3.1.1 Common-cause variation and special-cause variation

A cornerstone in the quality discourse is proper understanding, analysis and treatment of variation (Deming 2000, Bergman and Klefsjö 2003). To be able to predict a system process and act to improve it, we need to understand the causes of variation in the process. The first ideas and analysis approaches in this direction came from Walter A. Shewhart (see Shewhart 1931, 1939). He was concerned about predictability and argued that there are some systems for which the outcome variation may be described as coming from a constant system of chance causes, while other systems have causes of variation which do not give a predictable pattern of variation – they have what he called "assignable causes of variation" (Bergman 2009). It is common today to refer to these two different causes of variation as common causes and special causes, respectively (Deming 2000). A system with only chance (common) causes is considered to be under statistical control and its outcome is "predictable within limits". Following Shewhart, the special causes are identified by a Control Chart, where limits are defined for when the process is and is not considered to be under control (typically set as +/– 3 standard deviations from the mean); see Figure 3.1. If such limits are defined by one and seven leakages respectively in one month, the observations indicate that the system is under control. If we had observed (say) nine leakages in one month, the system would have been out of control – a special cause being the explanation.

If we study the quality management literature today, we will find that there are many different ways of defining the concepts of common- and special-cause variation. They are all related to the ideas of Shewhart, but the variation is quite large. Here are some examples:

Examples of common-cause definitions

c1) variation belonging to the system (the responsibility of the management) (Deming 2000)
c2) variation manifested by causes found to be inherent in a process (Cottman 1993)
c3) random variation that is inherent in the process and impossible to eliminate, unless we change the very design, the process or product (Magnussen *et al.* 2003)
c4) variation is considered to be due to the inherent nature of the process and cannot be altered without changing the process itself (Woodall 2000)
c5) the natural inherent variation in the process output when all input variables remain stable, that is, independent and identically distributed (Majeske and Hammett 2003)

Examples of special-cause definitions

s1) variation not belonging to the system (Deming 2000)
s2) variation that occurs due to a significant outside source (Cottman 1993)
s3) variation due to causes that are non-random, relatively few, and over time or in effect give unpredictable and relatively large contribution to variation (Magnussen *et al.* 2003)
s4) unusual shocks or other disruptions to the process, the causes of which can and should be removed (Woodall 2000)
s5) any increase in product variability above the level of common-cause variation (Majeske and Hammett 2003)

However, if we look at various internet websites on quality management we find a number of other definitions, for example (see e.g. Wikipedia 2013):

Common-cause definitions

c6) variation that arises from the system process and the way the process is organised and operated
c7) variation that arises due to phenomena that are constantly active within the system
c8) variation that reflects a historical experience base

Special-cause definitions

s6) variation that arises from phenomena that are new, unanticipated or emergent within the system
s7) variation that is caused by some inherent change in the system or our knowledge of it
s8) variation outside the historical experience base

Although this review shows that there is no consensus on how to define and understand the notions of common- and special-cause variation, it is not difficult to see the contours of the underlying ideas in the concepts: common causes capture 'normal' system variation, whereas special causes are linked to unusual variation and surprises. The term 'risk' in this context is commonly used in relation to missed alarms and false alarms and the choice of control limits; the perspective adopted is a standard statistical one. A natural issue to explore in this context is how the concepts and ideas of the quality discourse are linked to those of risk assessment – this link is the core of the scope of the coming analysis. Also the risk assessment field to a large extent concerns variation, uncertainties, the unforeseen and surprises, which all characterise the two variation concepts of the quality discourse, but its perspective extends beyond the statistical. The focus is on the consequences of the activities, and related probabilities and uncertainties. Through a thorough analysis of the meaning of common-cause and special-cause variation, the aim is to identify ideas and concepts that also have relevance for risk assessment and management. We will argue that common-cause variation is linked to aleatory uncertainties in risk assessment contexts, and that there are epistemic uncertainties about the common-cause variation, the special-cause variation and the model output compared to the real variation (uncertainty about the model error).

Clarifications of the meaning of common- and special-cause variation

We will use our two examples to illustrate the discussion: the die game and leakages on offshore installations.

DIE EXAMPLE

The above review presented a number of definitions of the concepts of common causes and special causes. Now we will look more closely at these two concepts. Let us start with the simplest case – our die-throwing game. Clearly, if the die is a standard/ideal one, the variation will be given by a chance/frequentist probability distribution p_i, i = 1, 2, ..., 6, where p_i = 1/6 is the fraction of times the die shows *i* when being thrown over and over again a large number of times (in theory, an infinite number of times, assuming the die does not wear out over time). This variation is due to common causes. All the definitions of common causes referred to above match the chance distribution p_i.

Special causes in this case could be a replacement of the die with a non-normal die so that the chance distribution is changed (for shorter or longer periods of time), deterioration of the die so that the chance distribution is changed, and so on.

For this simple case the terms are well-defined and easy to understand. Common causes are linked to the variation modelled by the chance distribution, the probability model, and the special causes are linked to variation as a result of changes in the phenomena justifying the established probability model.

Now suppose we observe the outcomes from some throws of a die having properties not known to us. Then we may ask: is the variation due to common causes or special causes? The answer, of course, is that we do not know. However, we can look at the data and even perform statistical testing, and if we have a large amount of data we should be able to see whether the p_i = 1/6 model is reasonable. With a limited amount of data, however, we must be careful, because we may experience large deviations from the p_i = 1/6 model. It could then be tempting to conclude that the variation is due to special causes, even though the die is fair. This is an example of a mistake of type (i) (Deming 2000), reacting to an outcome as if it were from a special cause, when actually it came from common-cause variation. A mistake of type (ii) would be to treat an outcome as if it were from common-cause variation, when actually it came from a special cause. For example, we could have a non-standard die that produces a variation distribution quite different from $p_i = 1/6$, but this would be difficult to identify from a limited amount of data alone.

The conclusion is that with few data it is difficult to discern what is common-cause variation and what is special-cause variation, without using other types of information and knowledge about the process generating the data. We must be careful in over-interpreting the numbers; the data alone can easily mislead us in many cases.

OFFSHORE INSTALLATION LEAKAGES

Let us leave the die example and gambling, and consider more typical real-world settings such as leakage data from the offshore installation referred to above. We have a process generating data $X_1, X_2, \ldots X_n$, but a probability model 'explaining' the data is not immediately available. As a first step we look at whether it is possible to argue for a specific distribution from the situation, for example a binomial distribution, a Poisson distribution or a normal distribution. The justification may be weaker than desired, but it is a fundamental feature of the quality tradition (see e.g. Deming 2000) to base quality management on theory, acknowledging that improved knowledge and learning strongly depend on having suitable models. The modelling may deviate considerably from the real world, but it can still be useful for decision-making. From the modelling we are able to develop limits for extreme outcomes (according to a Control Chart, as mentioned above), in order to identify special causes.

It should be noted that both Shewhart and Deming underline that no probability model is needed to define the limits of the Control Chart. The limits are chosen (typically representing +/– 3 standard deviations from the mean) because they are reasonable based on experience and from an economic point of view.

The distinction between common causes and special causes is context-dependent. We have to clarify what the common-cause variation covers and what it does not cover. There is no straightforward rule we can apply here. Consider the example of gas leakages in Figure 3.1. A number of causes of leakages have been identified (Røed *et al.* 2012), linked to six main categories as shown in Table 3.1. The interesting question here is: what are the common causes and what are the special causes?

The answer is that this has to be discussed in the concrete case, but one possible solution is that we consider all these causes to be common causes. The variation seen in Figure 3.1 is due to leakages from any of these causes. Given the present design and organisation, there is stability in the number of leakages, as shown in Figure 3.1. To improve the level of leakages,

Table 3.1 Categories of leakage causes as defined by Røed *et al.* (2012)

Categories of leakage causes	Subcategories
A. Technical degradation of system	• degradation of valve sealing • degradation of flange gasket • loss of bolt tensioning • fatigue • internal corrosion • external corrosion • other
B. Human intervention – introducing latent error	• incorrect binding/isolation • incorrect fitting of flanges or bolts during maintenance • valve(s) in incorrect position after maintenance • erroneous choice of installation of sealing device • inappropriate operation of valve(s) during manual operations • inappropriate operation of temporary hoses
C. Human intervention – causing immediate release	• breakdown of isolation system during maintenance (technical) • inappropriate operation of valve(s) during manual operation • work on wrong equipment (not known to be pressurised)
D. Process disturbance	• overpressure • overflow/overfilling
E. Inherent design errors	• design-related failures
F. External events	• impact from falling object • impact from bumping/collision

fundamental changes need to be carried out concerning the design or at the organisational level. Such changes have been carried out on the Norwegian Continental Shelf in recent years, the result being a reduced number of such events (Vinnem 2012).

What then may be classified as special-cause variation in this case? It could, for example, be an increase in the number of leakages due to sabotage and criminal acts by a person involved with the operation of the installation. Another example could be the use of equipment from a supplier with many design errors, leading to significantly increased numbers of leakages. A key task of the quality management is to identify the special causes and eliminate them, and bring the process into a state of control. As Deming (2000) highlights, most variation is due to common causes (he often refers to a distribution of 94% common causes and 6% special causes).

The Poisson model represents the common causes as listed above (A to F in Table 3.1). If the null hypothesis is rejected, it may be due to a surprising number of events related to the common causes, or due to special causes. We cannot know from the data. The significance level is computed on the basis of the Poisson model with rate 4, and hence we must be careful when interpreting this level. Say the significance level is 5%. Then, in one out of twenty cases we would conclude wrongly that the process is out of control. However, this line of argument does not take into account the special causes, and is, according to Shewhart and Deming (Deming 2000, p. 176), an argument for a pragmatic approach on how to determine the limits, as mentioned above.

Following a Bayesian approach, we would compute a posterior distribution of λ given the observations of the leakage numbers. Here, the Poisson model also provides the analysis basis, with the observations viewed as independent given λ and exchangeable unconditionally. In fact, Shewhart (1939) used exchangeability as a criterion for a process to be in control (Bergman 2009) – a sequence of random quantities is exchangeable if their joint probability distributions are independent of the order of the quantities in the sequence. In a Bayesian framework it is also common to define loss functions and search for optimal decisions using a suitable criterion, typically the expected loss.

In the die example it is easy to understand what the model is describing, but in the leakage case it is more difficult. The problem is that the model expresses the variation we would expect if we had an infinite number of months, whereas the data are limited to a few months. Clearly, the infinite case is based on thought constructions of how the system would behave if we could repeat the months under similar conditions – or in other words, if we could think about an infinite number of similar installations. 'Similar' here refers to the same type of equipment, the same operational procedures, the same type of personnel position and so on, but allowing for variation in design and organisational aspects, for example linked to equipment quality and human performance. However, giving precise specifications for an infinite population of such similar months is not easy, as discussed in Section 1.1; see also Aven (2012a, p. 37).

The modelling must be considered a tool for gaining insights, not for producing exact calculations and probabilities. It may be hard to justify a specific probability model and fully understand what type of variation it reflects. Nevertheless, the approach can produce valuable decision support for risk management. It requires a sufficient amount of data to be generated, which could lead to the use of indicators that are more or less interesting for making judgements about the overall performance of the system and risks. In the leakage case, the requirement for a large amount of data leads to a focus on small leakages, which make a lower risk contribution than those of large leakages. The larger leakages represent the biggest risk, but there are few such events and they alone cannot therefore provide the basis for the variation analysis.

The link to risk and unforeseen and surprising events in risk assessment and management

Using the two examples from the above analysis as illustrations, this section discusses how common-cause variation and special-cause variation relate to fundamental concepts in risk assessment and management, including the concept of the black swan.

Let us first reconsider the die example and place it in a risk context. Let us assume that the outcome 1 is an undesirable state or outcome, whereas the other outcomes are desirable. In the first situation studied, the chance distribution for the die was known to be $p_i = 1/6$, the distribution reflecting common-cause failures. Restricting attention to the undesirable outcome 1, we would express risk by the chance $p_1 = 1/6$, in line with the common perspective on risk, seeing risk as the pair of consequences (loss) and probabilities.

As the chance of interest is known, it follows that the magnitude of the risk is also known. Of course the probability 1/6 is conditional on the assumption that the die is fair, but this risk perspective does not prescribe how to add the 'value' of this assumption to the risk description. In practice we often see that it is not added at all, for example when using the result of risk assessment to compare computed probabilities with predefined acceptance limits; see Section 5.1.2. Following the quality management discourse as explained above, it is stressed that the probability refers to just one type of variation and that we also need to pay attention to special-cause variation. We can read this as stating that there is also a risk contribution from this type of variation, although the chance does not reflect it.

It is of course possible to redefine risk to also cover special-cause variation, within the perspective that risk is given by the pair of consequences and probability. However, we need to be careful in how we understand the probability dimension. Special causes could make it not meaningful to speak about chances reflecting variation: the existence of a long-run fraction of throws with outcome 1, the exchangeability judgement, may not be justified. For a specific future throw, we can always assign subjective probabilities and let risk be described by this number, but with this approach we lose the power of the statistical machinery using a probability model. The implication in a practical setting is that one sticks to the assumption that the process is under control and the probability model is applicable. The special causes are to be treated outside this framework. Unfortunately, in many situations where risk is studied, little attention is paid to the special-cause variation aspect (Aven 2014a); the focus is on the computed probabilities. With the quality discourse perspective added, the special causes – the surprise dimension of risk – will be an integrated part of the risk analysis.

This discussion motivates a risk perspective that sees beyond probability, as recommended in this book.

Let us say that the special-cause variation is restricted to only changing the chance distribution. Then, an adjusted probability model can be defined by chances p_i as above, but now these parameters are unknown. There are epistemic uncertainties about what the true values of these parameters are. We have moved to the case studied above where the die's properties are not known to us. Then we are in the common framework studied in risk assessment, where we have the distinction between aleatory uncertainties reflecting variation in defined populations represented by probability models, and epistemic uncertainties reflecting the analysts' lack of knowledge about unknown quantities, for example the values of the parameters of the probability models (Apostolakis 1990). The strength of this framework is that new information can be easily incorporated and the epistemic uncertainty assessments can be updated (normally using Bayes' formula). The problem is the fixed model. As stated by Bergman (2009) (see also Aven and Bergman 2012): "The problem is the closeness of the approach – once the model is chosen, it does not seem to be possible to escape – whatever we observe, we cannot escape." The common-cause variation, which is modelled by the probability model, does not cover all types of variations. There are risk contributors which are not included. Compared to the knowledge provided by the common-cause variation, a special-cause event would come as a surprise. It is thus a black swan if we adopt the terminology used in Sections 1.7 and 3.4, provided the consequences are judged to be extreme.

To summarise, common-cause variation is modelled by a probability model, for which there can be epistemic uncertainty about the model output variation compared to the real variation (uncertainty about the model error, Aven and Zio [2013]). This uncertainty is due to lack of knowledge about the common-cause variation as well as the special-cause variation, as illustrated by Figure 3.2. Including the special-cause variation ensures the completeness of risk contributors and is thus to be seen as a tool for treating completeness uncertainty, as discussed in the risk literature (see e.g. NUREG 2009).

The situation is similar for the leakage example. The special-cause variations will cause the Poisson model to produce poor predictions; this could lead to extreme outcomes in future months, with a high number of leakages compared to the predictions provided by the common-cause variation and the Poisson model. From a risk management point of view, the challenge is to use the data to identify the special causes quickly and rectify them. However, proper interpretation is not always straightforward. As mentioned above, when confronting the variation we can make two main mistakes (Deming 2000): (i) reacting to an outcome as if it were from a special cause, when actually it came from common causes of variation; and (ii) treating an outcome as if it were from common causes of variation, when actually it came from a special cause. For example, we may experience quite a high number of leakages in one month due to many independent random events, but we might believe that this increase is due to a special cause, for example systematic violations of procedures. Alternatively, we may

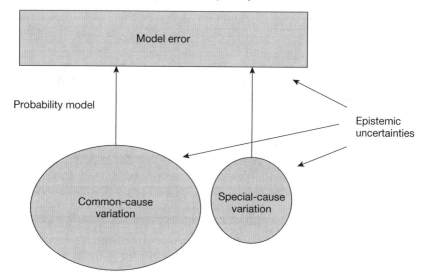

Figure 3.2 Illustration of contributors to model error (difference between the model output variation compared to the real variation) (based on Aven 2014a)

experience quite a high number of leakages due to defects in all the equipment used for some types of operations, but believe that this increase is due to random events and consequently not take action until a large number of events have occurred.

Mistakes will be more difficult to discern in this case. However, by highlighting the two different types of causes, management obtains a procedure for how to react:

- If the indicator exceeds the limit, look for special causes and take action to try to remove them.
- If the indicator does not exceed the limit, treat the variation as common-cause variation and place the need for rectification in a wider management and judgement process, considering the total variation and the need for a fundamental improvement of the system.

Mistakes will be made using such an approach, but this is unavoidable. The quality tradition has provided ample evidence that the focus on variation leads to high performance in a long-term perspective.

Let us return to the pickpocketing case on the Paris Metro (Section 1.1). I decided to also allow for peak periods to be part of the normal variation – since I need, for various reasons, to take the train at such points in time. A set of measures as indicated in Section 1.1 was implemented.

TECHNICAL QUANTITATIVE RISK ASSESSMENT IN THE PLANNING STAGE

In the above discussion we focused on situations where risk indicators are available. Now let us consider a system at the planning stage, and study the risk in relation to events that could occur and might lead to fatalities. A common tool for analysing such risk is event tree analysis. The question here is whether the distinction between common causes and special causes also has a meaning in this situation, and how it may strengthen the risk assessment process.

Think of the event tree in relation to a specific initiating event, for example a major gas leakage. The tree, with chances for the various events in the tree (the initiating event and branching events), constitutes the probability model reflecting common-cause variation. The variation is not actually observed, as the system is considered in the planning phase, but the concept can be used as a thought construction. Thinking of a large number of realisations of the system, we would get variations according to this probability model. We are back to the situations considered in the above two cases and in Figure 3.2. We need to address the ability of the probability model to represent common-cause variation, as well as the influence of the special-cause variation on the model error (the difference between the model output variation and the real variation).

Alternatively, we could proceed without introducing a probability model in this case (as demonstrated by Aven 2012d). We simply use subjective probabilities to express the epistemic uncertainties related to the specific unknown future events. The uncertainty assessments and associated probability assignments reflect common causes, although they are not modelled using probability models. Special causes are treated separately, for example as indicated in Aven (2013b), by addressing the uncertainties concealed in the assumptions on which the probabilities are based, and by addressing potential surprises relative to the knowledge and beliefs of the analysts (black swans). Of course, this approach can also be used for the case when probability models are introduced. It is based on a concept of risk which extends beyond probabilities. The focus is on uncertainties, knowledge and surprises, and includes both the common-cause and special-cause variation. Risk captures both types of variation, although to a large extent prevailing thinking and practice see risk as associated with common-cause variation. The quality discourse's highlighting of special causes coincides with the trend we have seen in recent years to include the surprise dimension in risk conceptualisation and assessments.

As mentioned in Section 2.2.3, it is common in the economic literature to distinguish between risk and uncertainty, based on the availability of information and knowledge. In the case of risk, the probability distribution of the quantities studied can be determined objectively, whereas in the case of uncertainty these probabilities must be estimated or assigned on a subjective basis. We observe that in the case where common-cause variation is known, the Knightian risk coincides with this variation.

Conclusions

In this section we have discussed how the key concepts used in quality management, common-cause variation and special-cause variation, are linked to fundamental concepts in the fields of risk assessment and management. We have argued that common-cause variation is linked to aleatory uncertainties in risk assessment contexts, and that there are epistemic uncertainties about common-cause variation, special-cause variation and model output compared to real variation (uncertainty about the model error). Highlighting the special-cause variation concept in risk assessment can lead to a stronger focus on potential surprises, which is necessary because the common risk assessment practice pays little attention to this aspect. To be able to adequately reflect both types of variation, a suitable broad risk concept and framework are needed, as discussed above.

3.2 Knowledge

It is common to distinguish between three types of knowledge: know-how (skill), know-that (propositional knowledge), and acquaintance knowledge (Lemos 2007). 'Knowing how to ride a bike' is an example of know-how, the statement 'I know that Oslo is the capital of Norway' is an example of propositional knowledge, while 'I know John' is an instance of acquaintance knowledge. In this book our main interest is in propositional knowledge, but also in aspects of know-how.

The traditional perspective states that (propositional) knowledge is made up of justified true beliefs. This meaning of knowledge is the point of departure for most text books on knowledge. However, this definition may be challenged, and in this book we consider knowledge to be justified beliefs. A risk analysis group may have strong knowledge about how a system works and may be able to provide strong arguments in favour of why it will not fail over the next year, but the group cannot know for sure whether or not it will in fact fail. Nobody can. However, the group's beliefs can be expressed through a probability. As another example, consider a case where a group of experts believe that a system will not be able to withstand a specific load. Their belief is based on data and information, modelling and analysis, but they may be wrong. The system may be able to withstand this load. As a third example, is it true that the (frequentist) probability of a fatal accident in a process plant is higher than $1 \cdot 10^{-4}$, even if a large amount of data shows this and all experts agree on it? Leaving aside the issue that such probabilities may be difficult to define, we cannot say that we have a true belief that this probability is so large, because the probability is unknown. However, we may have a strong belief, and we can even introduce (subjective) probabilities to express the strength of the belief.

Following this line of thinking, clearly knowledge cannot be objective, since a belief is expressed by a person. In general, therefore, knowledge needs to be considered as subjective or at best inter-subjective among people, for example experts.

From such a perspective the term 'justified' becomes critical. Philosophers and others have discussed this issue since ancient times. In this book, justifiability is linked to being a result of a reliable process, a process that generally produces true beliefs. It applies to the justification of a specific statement by an individual as well as broad justifications of scientific theses. In line with Hansson (2013), the perspective taken here is that science is a means to produce knowledge in this sense:

> Science (in the broad sense) is the practice that provides us with the most reliable (i.e. epistemically most warranted) statements that can be made, at the time being, on subject matter covered by the community of knowledge disciplines, i.e. on nature, ourselves as human beings, our societies, our physical constructions, and our thought constructions.
>
> (Hansson 2013)

In addition, criteria such as scientific fruitfulness and explanatory power need to be applied to determine scientific quality (Hansson and Aven 2014).

In this section we will look more closely into such reliable processes for three cases: the assignment of a knowledge-based probability, quantified risk assessments, and finally the science of risk analysis. We will see how our perspective allows for and stimulates scrutiny of the quality of the reliability process. The degree of reliability will always be an issue for debate. We will argue that for many types of situations where the issue is knowledge, it is useful to distinguish between the available data (D), information (I), knowledge (in the sense of know-how and justified beliefs) (K) and wisdom (W), i.e. the various elements of the so-called DIKW hierarchy as mentioned in the introduction to this chapter.

The DIKW hierarchy constitutes a fundamental framework for analysing knowledge (interpreted in a wide sense) in many types of applications. A number of extensions and refinements of the DIKW hierarchy have been presented since the 1980s, and there has been a continuous discussion about the elements and structure of the hierarchy. Just as interpretations are required for all conceptual structures, so it is the case for the DIKW model. Aspects of the DIKW hierarchy are controversial – see e.g. Frické (2009) – but for the purpose of the present book we simply acknowledge the hierarchy as an interesting model for distinguishing between the various components of knowledge (understood in a broad sense). Through our analysis we will make specific interpretations for the DIKW elements adapted to the risk context we are addressing.

Although there are differences in perspectives and views about DIKW and its properties, there is a central core. This may be summarised briefly as follows:

Data:

- Data are the symbolic representation of observable properties of the world (Rowley 2007).
- Data are symbols that represent the properties of objects, events and their environments. They are products of observation (Ackoff 1989, p. 3).

  Example: a file comprising the (raw) numbers which refer to the observed events in different periods of time.

Information:

- Relevant, or usable, or significant, or meaningful, or processed, data (Rowley 2007).
- Information systems generate, store, retrieve, and process data. In many cases their processing is statistical or arithmetical. In either case, information is inferred from data (Ackoff 1989).

Information can be inferred from data on the number of events of a specific type, for example the average value, with some illustrations showing trends over time etc. The idea is of a human asking a question beginning with "who", "what", "where", "when", or "how many" (Ackoff 1989, p. 3), and the data is processed into an answer (Frické 2009). When this happens the data becomes 'information'. Data itself is of no value until it is transformed into a relevant form.

Knowledge:

Here interpreted as know-how and propositional knowledge (understood as justified beliefs), as discussed above. The justified beliefs are often formulated as assumptions.

Wisdom:

There has not been such an intense focus on this term as on the DIK elements, but it has been more thoroughly discussed in recent years; see for example Zeleny (2006) and Rowley (2006, 2007). An example of the many definitions suggested is the following by Rowley (2006, p. 257):

    wisdom is the capacity to put into action the most appropriate behaviour, taking into account what is known (knowledge) and what does the most good (ethical and social considerations).

Zeleny (2006) writes that wisdom is not about "Because I can", "Because it is there" or "Because I must" – these are the traditional explanations of the unwise. Many informed people know what to do, quite a few knowledgeable experts know how to do it, but only a few wise persons know and can fully explicate why it should be done. In line with these ideas the following metaphor applies: data is "know-nothing", information is "know-what", knowledge is "know-how", and wisdom is "know-why" (Zeleny 2006).

3.2.1 Assigning a knowledge-based probability

A person, let us call her Susan, needs to assign a knowledge-based (subjective) probability P(A|K), where K is Susan's background knowledge for this assignment. The issue we raise here is: what type of knowledge is covered by K?

To provide a response to this question, the DIKW hierarchy is useful. Clearly K may comprise data, information and justified beliefs. Consider a case where John has the following failure data available for a unit: 100 units, 5 failures. The observed failure rate is thus 0.05. The analyst argues that these data are relevant for future observations (her justified belief is the assumption that the unit considered is similar to those observed) and she assigns a probability of 0.05 for a future unit to fail.

Let us complicate the situation somewhat and assume that no relevant unit data exists. Susan has to rely on experts to provide judgements about the likelihood that the unit will fail, or perhaps on data from units not directly comparable to the one studied. The data in this case can be the judgement of the experts, information in the form of some aggregated figures, and justified beliefs about how this unit compares to those used for generating D and I. In addition, there are a number of know-how elements: if Susan is a well-trained probability assigner, she knows how to combine expert judgements and hard data, how to interpret the produced probability, etc. We may also add that she should then be able to present the probability in the right way in the context used, for example reflecting the limitation of the approach, the fact that there are other perspectives than the one used, etc. However, it is more natural to categorise this feature as wisdom, since it requires insights that are beyond those of most risk analysts today. We will discuss this in more depth in the next section when we are looking into quantified risk analysis.

3.2.2 Quantified risk assessment

When analysing risk, statisticians introduce a probability model with parameters, for example a Poisson model with occurrence rate λ. If N(t) denotes the number of accident events in the interval [0,t], this means that

$$P_f(N(t) = x) = (\lambda t)^x e^{-\lambda t}/x!, \quad x = 0, 1, 2, \ldots, \tag{3.1}$$

where λ equals the expected number of accident events per unit of time, i.e. $\lambda = E_f[N(t)/t]$. Here P_f is a frequentist probability, understood as the fraction of time in which the event of interest occurs if the situation considered could be repeated infinitely under similar conditions. The E_f is understood as the expectation with respect to this frequentist probability, and is interpreted as the average number of occurrences per unit of time when repeating the situation considered over and over again infinitely. A frequentist probability reflects the so-called stochastic (aleatory) uncertainty, i.e. the variation in the infinite population considered. Based on observations of the process $N(t)$, an estimate of the occurrence rate λ is produced. For example, if $n(t)$ is the observed number of accident events in the interval $[0,t]$, the common estimate for λ is $\lambda^* = n(t)/t$.

In this context, risk is commonly defined by the combination of $N(t)$ and the associated frequentist probability given by formula (3.1), which we schematically, and in line with the standard terminology used in Section 2.2, refer to as 'Risk = (C,P_f)', where $C = N(t)$ is the consequence of the activity considered.

Because this risk is unknown it has to be estimated, and we are led to a risk description (C,P_f^*), where * indicates that the probability is estimated (by replacing λ by λ^* in [3.1]).

The issue we discuss in the following is how this risk perspective and others are linked to various features of knowledge (interpreting this term in a broad sense). We question the degree to which the risk perspective is able to capture the various elements of the DIKW hierarchy. For example, for the (C,P_f^*) perspective we can identify the following set of DIKW components:

Observation $n(t)$:	Data (D)
Estimates λ^* and P_f^*:	Information (I) (from the analyst to the decision-maker)
Construction of the probability model (3.1):	Knowledge (K) (analyst knowledge)

When the author of this book worked with this topic, an initial hypothesis was that prevailing risk perspectives do not adequately incorporate the wisdom aspect which highlights reflections, for example on the difference between common practice and established knowledge (which may lead to use of a specific probability model) on the one hand, and the future and the potential for surprises on the other.

In line with the risk concepts discussed in Section 2.2, we distinguish between three main risk perspectives (see Table 3.2): the traditional statistical approach, the traditional Bayesian approach, and the (C,U) perspective.

Table 3.2 Characterisations of three perspectives on risk, BK: background knowledge (based on Aven 2013b)

Approach	Risk	Uncertainty representation	Risk description	Perspective on analysing data and information
Traditional statistical approach	(A,C,P_f)	Frequentist probabilities	(A',C',P_f^*,CI) A', C': specific events and consequences CI: confidence interval	Traditional statistical analysis using probability models, point estimates, confidence intervals and hypothesis testing
Traditional Bayesian approach	(A,C,P_f), (A,C,P) or more precisely (A',C',P,BK) A', C': specific events and consequences BK: background knowledge that P is based on	Subjective probabilities (also referred to as judgemental or knowledge-based probabilities) Frequentist probabilities (also referred to as chances in this setting) are used as a tool – as parameters of probability models to support the assignment process of the subjective probabilities	(A',C',P,BK)	Standard Bayesian analysis using probability models, Bayesian updating based on prior distributions and the application of Bayes' formula to produce the posterior distribution, and finally, the predictive distributions of the quantities of interest and the observables are derived by applying the law of total probability
General risk approach (based on events/consequences and uncertainties)	(A,C,U)	Any representation (measure) of uncertainty Q, for example P or imprecise probabilities	(A',C',Q,BK)	The Bayesian approach is used as a tool when appropriate, but it is acknowledged that there is a need to see beyond this tool in many cases to reflect situations characterised by large uncertainties and the potential for surprises

Tables 3.3 to 3.5 show a structure matching the key DIKW elements and the three risk approaches described above, using the Poisson example introduced earlier as an illustration. Schematically the structure expresses that:

Data = the input to risk assessment.

Information = the risk description.

Knowledge (for the decision-maker) = understanding the risk description (which means grasping or comprehending the meaning intended or expressed by the risk description).

Knowledge (for analysts) = understanding how to perform the risk assessment and understanding the risk description.

Wisdom (for the decision-maker) = the ability to use the results of the analysis in the right way.

Wisdom (for analysts) = the ability to present the results of the analysis in the right way.

The data (D) element for the traditional statistical approach covers hard data only, whereas the Bayesian approach and the general risk approach also include expert judgements, for example an expert expressing his/her judgements about the value of unknown quantities (expressed by a subjective probability distribution, for instance). The data are used to provide information, which we interpret as the description of risk in this context. Hence the information goes from analysts to the decision-maker (and potentially other stakeholders). The decision-maker, or any other user of the risk assessment, is informed by the risk description. In the traditional statistical approach, the risk description covers estimates of the parameters and confidence intervals, linked with explanations of what the quantities express. In the Bayesian approach, subjective probability distributions of the quantities of interest are also provided. For the general risk approach, the risk description is based on the characterisation made by the uncertainty measure Q, which could be probability or another representation of uncertainty, for example imprecise probability. For example, the risk analysis team may express that $0.10 \leq P(\lambda > \lambda_0) \leq 0.50$, where λ_0 is a specified number and P is a subjective probability; the analysis team is not willing to make a more precise assignment. Or, the analysis team could use precise probabilities with scores for the importance of the assumptions on which the probabilities are based, as outlined in Section 4.2. The team may also limit itself to a qualitative description. For all approaches, the assumptions on which the analysis is based need to be reported.

When we come to the knowledge and wisdom elements, we need to distinguish between the perspective of the analysts and that of the decision-

Table 3.3 The DIKW elements linked with the risk perspectives defined in Table 3.2, using the Poisson example as an illustration (based on Aven 2013b)

Approach *DIKW elements*	*Traditional statistical approach*	*Traditional Bayesian approach*	*General risk approach (based on events/consequences and uncertainties)*
Data	Observational data n(t)	Observational data n(t), expert judgements	Observational data n(t), expert judgements
Information (from the risk analysts to the decision-maker)	Estimates of parameters, with explanations of what the numbers mean Assumptions on which the analysis is based λ^* and P_f^*, CI	Estimates of parameters, subjective probability distributions, predictions, with explanations of what the numbers mean Assumptions on which the analysis is based λ^*, P(C'≤c), E[C'], C^*, ...	Estimates, predictions, assigned values of Q, ..., with explanations of what the description/quantities mean Assumptions on which the analysis is based C^*, Q(A'), Q(C'≤c), ...

maker. First, let us look at these elements from the decision-maker's perspective (see Table 3.4).

For all three risk approaches, knowledge is about having an understanding of the potential hazards/threats, their potential consequences and how professional risk analysts judge the overall risk. It is about understanding the result of the risk assessments, what the overall approach is able to do and what its limitations are. The decision-maker will place the presented risk description in the context of his/her own reference system, for example reflecting his/her degree of confidence in the risk analysts and the results obtained. If a 'true' risk is presumed to exist, as in the traditional statistical approach (there are presumed to be some correct frequentist probabilities that we seek to accurately estimate), knowledge is also about having an understanding of the 'true' risk for the activity considered. For the other two risk approaches, the focus is on having knowledge about the risk phenomena – for example, being able to understand how a hazardous event can lead to some severe consequences, as mentioned at the beginning of this paragraph.

As shown in Table 3.4, the wisdom element W is interpreted along the path mentioned in the introduction of this Section 3.2: the capacity to put into action the most appropriate behaviour – here, being able to use the results of the analysis in the right way in a decision-making context, reflecting the capabilities and limitations of the approach used. To make adequate use of the results of risk assessments, ethical and social considerations are required.

Table 3.4 The KW elements linked with the risk perspectives defined in Table 3.2, using the Poisson example as an illustration – the decision-maker perspective (based on Aven 2013b)

Approach DIKW elements	Traditional statistical approach	Traditional Bayesian approach	General risk approach (based on events/ consequences and uncertainties)
Knowledge (Decision-maker)	Having a good picture of what the "true" risk is for the activity considered Having a good understanding of the potential hazards/threats, their potential consequences and how professional risk analysts judge the overall risk Understanding the result of the risk assessments, "what the overall approach is able to do and what its limitations are" (the text " " is hereafter referred to as (3.2))	Having a good understanding of the potential hazards/threats, their potential consequences and how professional risk analysts judge the overall risk Understanding the result of the risk assessments, what the overall approach is able to do and what its limitations are (3.2)	Having a good understanding of the potential hazards/threats, their potential consequences and how professional risk analysts judge the overall risk Understanding the result of the risk assessments, what the overall approach is able to do and what its limitations are (3.2)
Wisdom (Decision-maker and other stakeholders)	Being able to use the results of the analysis in the right way reflecting a decision-making context inter alia (3.2)	Being able to use the results of the analysis in the right way in a decision-making context reflecting inter alia (3.2)	Being able to use the results of the analysis in the right way reflecting a decision-making context inter alia (3.2)

The alternative path mentioned above. The know-why feature – understanding why the approach taken is appropriate in a specific case – is placed under the knowledge element K. The proper placing may of course be discussed; decisive for the choice made here was the view that there are strong know-that aspects related to why a particular approach is suitable in a specific case. For example, if 'hard data' are available and are the only data source allowed, in addition to modelling, the traditional statistical approach is the appropriate one. However, if it is important to highlight potential surprises concealed by the probabilistic analysis, the general risk approach is the one to choose.

Finally, let us consider the analysts' perspective on knowledge and wisdom (Table 3.5). We interpret the knowledge element K as knowing how to perform the risk assessment and knowing how to understand the risk description with all the quantities introduced and reported, i.e. the know-how (skill) and the know-that of propositional knowledge, as mentioned at the beginning of Section 3.2. Thus, if a subjective probability is introduced and used, the analysts must know how to interpret this measure (as also noted in Section 3.2.1). Therefore, knowledge means that the analysts know what the overall approach is able to do and what its limitations are. For instance, if the traditional statistical approach is adopted, the analysts must know what a confidence interval expresses and how to compute this interval in a practical case. If a specific probability model is adopted, the analysts must know that this model represents an accurate representation of the phenomena studied. If the general risk approach is adopted, the analysts needs to know inter alia which measure Q to use and how to interpret this measure.

The wisdom element for the analysts is interpreted as being able to present the results of the analysis in the right way in a decision-making context, paying due attention to the limitations of the approach and the existence of alternative approaches that may provide different results than the one used.

It has been commented that the wisdom element of the analysts is closely linked to the wisdom element of the decision-maker; what is the 'right way' to present the results of a risk analysis? Is it not the way that will convince the decision-maker to make a good decision? (Aven 2013b).

Clearly there is a link, but care has to be taken when integrating these perspectives. The risk analysts can perform a professional and wise interpretation of the results of the analysis, without in any way giving recommendations for what decision alternative to choose: the analysts' role is to provide decision support, nothing more. The issue relates to the well-known distinction between 'evidence' and 'values' in risk management; see discussion in Section 5.1.

Discussion

A main aim of a structure linking the DIKW elements and the three risk perspectives presented in the previous section has been to reveal and point

Table 3.5 The KW elements linked with the risk perspectives defined in Table 3.2 using the Poisson example as an illustration – the analyst perspective (based on Aven 2013b)

Approach / DIKW elements	Traditional statistical approach	Traditional Bayesian approach	General risk approach (based on events/consequences and uncertainties)
Knowledge (Analysts)	Having a good understanding of the risk description and its features, for example related to potential hazards/threats and their potential consequences. Construction of probability models – Poisson model (3.1) How to perform the estimation How to understand the results of the estimation – for example, what the confidence interval expresses Understand what the overall approach is able to do and what its limitations are (3.2) A key issue related to (3.2) is the fact that the approach does not describe epistemic uncertainties Understand why this approach is appropriate in a specific case	Having a good understanding of the risk description and its features, for example related to potential hazards/threats and their potential consequences. Construction of probability models – Poisson model (3.1) How to perform the estimation and the subjective probability assignments integrating 'hard data' (like n(t)) and expert judgements How to understand the produced quantities – for example a subjective probability and credibility intervals Understand what the overall approach is able to do and what its limitations are (3.2) A key issue related to (3.2) is the fact that the approach is based on subjective probabilities which can conceal uncertainties in the background knowledge BK Understand why this approach is appropriate in a specific case	Having a good understanding of the risk description and its features, for example related to potential hazards/threats and their potential consequences. Construction of probability models if introduced – Poisson model (3.1) How to perform the estimation/predictions and the assignment of Q How to understand the produced quantities – for example Q(A') Understand what the overall approach is able to do and what its limitations are (3.2) A key issue related to (3.2) is the fact that the actual A need not be captured by the set of A's Understand why this approach is appropriate in a specific case
Wisdom (Analysts)	Being able to present the results of the analysis in the right way in a decision-making context, reflecting for example (3.2) and that there exist other perspectives than the one used	Being able to present the results of the analysis in the right way in a decision-making context, reflecting for example (3.2) and that there exist other perspectives than the one used	Being able to present the results of the analysis in the right way in a decision-making context, reflecting for example (3.2) and that there exist other perspectives than the one used

to differences between these perspectives. Some of these fundamental differences are shown in Tables 3.3 to 3.5, while others are hidden in the general formulations. For example, consider the statement: "Understand what the overall approach is able to do and what its limitations are" (referred to as (3.2) in Tables 3.4 and 3.5). Clearly, a full description of this point would show large differences. For example, the traditional statistical approach is only relevant for situations where a probability model can be justified, where it includes the frequentist probabilities P_f, and where a large amount of relevant data exists. Similarly, we can identify frame conditions for the two other approaches. A key issue for the Bayesian approach is the fact that it is based on subjective probabilities which can conceal uncertainties in the background knowledge BK (we write BK here so as not to confuse it with the K in the DIKW hierarchy).

A subjective probability (or more generally the measure Q) may be used to represent or describe the uncertainties of events and consequences. For example, we write P(A|BK) to show that this probability of an event A is conditional on the background knowledge (BK). As noted in Section 3.2.1, the BK essentially covers the DIK elements.

For all approaches we have to face the problem that the actual A need not be captured by the set of A's studied in the risk assessment. There is a potential for surprises – extreme events relative to the present knowledge. It is important to highlight such issues. Too often risk analysis results are presented without a context to indicate what the approach is actually doing and what its limitations are. Through the KW elements these issues are more focused, and this can contribute to an improved understanding of how best to use the different approaches as well as obtaining a better basis for selecting the proper approach in a specific case.

It is beyond the scope of the present book to provide 'operational tools' for how to implement the ideas discussed above. However, based on the above analysis it is quite straightforward to make a list of some key points that should be checked with regard to the execution of a risk assessment (further details may be checked in relation to each main type of approach and in specific cases, see also Chapter 4):

- Clarify which risk analysis approach is to be adopted and what the DIKW elements express.
- Clarify the meaning of all concepts introduced.
- Clarify what the various approaches and methods do and what their limitations are.
- See to it that all models introduced are properly justified.
- Present the results in a way that is in line with the overall analysis approach taken, give due consideration to the limitations of the approach, and adequately reflect the uncertainties.
- Use the results of the analysis as decision support (not to prescribe decisions), in a context where it is clear what the analysis does and what the limitations are.

All these points can be seen as fundamental quality assurance aspects, and as such one should expect that they are included in common practical procedures for risk assessment. However, this is not the case, as argued above (see also Aven 2011b). A main problem is that often the overall analysis approach is not clarified, and then it is difficult to follow up the other aspects because the necessary conceptual precision is lacking. The use of the DIKW elements can provide help for the analysts and the decision-maker to increase their precision level and help them address the right issues.

In the above analysis (Tables 3.3 to 3.5), the analysts' task is to inform the decision-maker. In practice there could be interactions – for example, when the decision-maker is involved in specifying a loss function of C' (the loss that is associated with specific values of C'). The question then is how to place the loss function in this structure: is it data?

Yes, the loss function should be added to the data row (Bayesian approach) in Table 3.3. In addition, the specification and use of the loss function is an important aspect in the knowledge and wisdom elements – for example, when stating: "Understand what the overall approach is able to do and what its limitations are" in Tables 3.4 and 3.5.

The above analysis is based on a set of common interpretations of the DIKW elements. However, as was stressed at the beginning of Section 3.2, there are other ways of interpreting the DIKW elements, and depending on the interpretation, we can arrive at different connections to risk. An example of an alternative view on the DIKW elements is provided by Hansson (2002). His thinking is interesting because he also discusses the DIK elements in relation to risk, although from a completely different perspective than the one presented above. For Hansson (2002), knowledge is defined as true, justified belief, and risk is placed in relation to knowledge as shown in Figure 3.3.

A key point in Hansson's analysis is that risk is a combination of knowledge and uncertainties. He writes: "When there is risk, there must be something that is unknown or has an unknown outcome, hence there must be uncertainty." This is in line with the (C,U) risk perspectives: A and C are not known, there are uncertainties about these quantities. Hansson continues: "For this uncertainty to constitute a risk for us, something must be known about it." The validity of this statement depends on the risk perspective. If risk = (A,C,P_f), there is some knowledge present because a frequentist probability has been justified, and if risk = (A',C',P,BK) there is some knowledge present because A' and C' have been identified. However, if the (A,C,U) risk perspective is considered, there is no need for knowledge to define risk beyond the fact that A and C are unknown. The underlying ideas of Figure 3.3 seem to be better represented by replacing 'risk' with 'risk description'. Then the risk of Figure 3.3 is close to specified consequences, uncertainty description and knowledge.

Hansson's definition of knowledge as *true* beliefs may be questioned. How can one know that a belief is true? Care should be shown in using

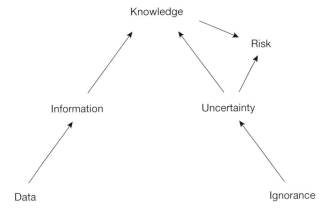

Figure 3.3 The place of 'risk' among epistemic concepts according to Hansson (2002) (Hansson emphasises that the intricate relations between risk and the other concepts are only hinted at by the placement of 'risk' in this figure)

terms like 'objectivity' and 'truth' in this context. Experts and others may agree on a specific statement, but then the proper wording is "agreement among experts, or broad inter-subjectivity if this can be achieved". Objectivity does not exist because the true underlying probabilities are not known. For risk assessment, it is important to acknowledge this. Analysts and agencies may have an interest in presenting an objective description of risk and may give the impression that they are able to do so, but their results are at best broadly inter-subjective.

When it comes to wisdom, there is clearly no objective answer: using the results of the analysis in the right way in a decision-making context means balancing different concerns and attributes, and people can perform the weighting differently. However, it still makes sense to refer to the wisdom concept because this balancing can be done in many ways, and wisdom can only be obtained if due attention is also given to the limitations of the approach.

Perhaps the main contribution of the above structure relates to the last two elements: knowledge and wisdom. Unfortunately, too much work on risk today is on data and information. Too seldom do we see that reflections are made on the justification of probability models, the relevance of data, and the potential for surprises compared to existing knowledge/beliefs. It is possible to conduct and use risk assessments based on the knowledge and wisdom elements in line with the above ideas and structures (Tables 3.3 to 3.5), for all three risk perspectives. However, history has shown that some risk perspectives seem to attract attitudes that are the opposite of balanced reflections, a search for finding and pronouncing the truth about risk. Such thinking cannot be justified and needs to be replaced by more humble approaches. The present analysis seeks to contribute to this end. The above reasoning disqualifies the use of any of the three perspectives, and in particular the traditional statistical approach and the Bayesian approach, unless proper weight is given to the knowledge and wisdom aspects. The general risk perspective has per definition a stronger motivation for seeing beyond the narrow analyses, but even here care must be taken – high quality requires all the features of the DIKW hierarchy to be incorporated.

3.2.3 The risk analysis science

In this section we see knowledge in relation to science, and in particular the science of risk analysis. We will argue that risk analysis is scientific in view of a set of commonly used definitions and criteria, which to a large extent are linked to knowledge. The key conclusion is that risk analysis (interpreted in a wide sense) is a scientific field of study when it is understood as consisting primarily of: (i) knowledge about risk-related phenomena, processes, events, etc.; and (ii) concepts, theories, frameworks, approaches, principles, methods and models to understand, assess, characterise, communicate and manage risk, in general and for specific applications (the instrumental part).

In many categorisations of scientific areas, such as research funding schemes, risk analysis is not mentioned. However, the subject has many scientific journals devoted to it, and there are quite a few university programmes on all levels, particularly in Europe and the US, covering risk assessment and risk management. From a sociological perspective it may be argued that risk analysis is a science since it has the standard characteristics of a science, including refereed journals and conferences, professorships and university programmes. However, for a principled discussion of its scientific status, the presence of such social practices is not sufficient. Instead, we have to focus on criteria of the types developed in the philosophy of science, and in particular whether risk analysis satisfies the normative criteria that justify the social institution of science. Like it or not, the social role of science is to provide us with reliable knowledge (suitably interpreted) on a wide variety of topics. In order to be a legitimate science, a discipline has to fulfil that function. Therefore, in order to determine whether risk analysis is scientific, we have to consider it from an epistemological point of view and investigate its methods and its claims to produce reliable knowledge.

Some of the earliest contributions to the discussion about risk analysis and science go back to Weinberg (1974) and the editorials of the first issue of the journal *Risk Analysis* (Cumming 1981, Weinberg 1981), in relation to the establishment of the Society of Risk Analysis. These authors conclude that the process of analysing or assessing risks involves science and consequently is a scientific activity, but risk assessment is not a scientific method per se; there are, and there will always be, strong trans-scientific elements in risk assessment. The term 'trans-scientific elements of risk assessments' means those questions that seem to be scientific but cannot yet be answered by science, for example predictions of rare events where the uncertainties are very large. Weinberg (1981) stresses that empirical scientific observation is inapplicable to the estimation of overall risk in the case of rare events, which are those instances where public policy most often demands assessment of risk. He refers to the intrinsic uncertainty reflected in the bitter controversy that rages over the safety of nuclear reactors.

3.2.3.1 What is science?

Science is a means to produce knowledge. Therefore, in order to characterise science, we need to specify its subject matter and the nature of the knowledge it provides on that subject matter. Furthermore, when science becomes involved in practical decision-making, as risk analysis invariably does, the impact of social norms and values on knowledge claims becomes a pivotal issue. Therefore, we also need to discuss how scientific knowledge relates to values and norms.

THE SUBJECT MATTER OF SCIENCE

Ideally, we would expect the delimitation of the subject matter of science to be based on fundamental epistemological criteria and therefore also to be the same in different languages. However, this is not the case. The English word 'science', with its counterparts in the Romance languages, covers a more limited group of disciplines than its translations into the other Germanic languages such as *Wissenschaft* (German), *wetenschap* (Dutch), *vitenskap* (Norwegian) and *vetenskap* (Swedish). Originally the word 'science' had a very broad meaning, covering nearly every type of knowledge or skill that is acquired through study, be it prosody or horse-riding. In the 1600s and 1700s the meaning of the term was restricted to systematic knowledge, and during the 1800s it was further restricted to denote the new, more empirical type of knowledge in the area previously called 'natural philosophy'. The word 'science' is still often used as a synonym of 'natural science', but it is also applied to some of the academic disciplines in the behavioural and social areas. Economics and sociology are often counted as sciences, whereas other academic disciplines such as those concerned with human history, arts and literature are not. *Wissenschaft* and its cognates in the other Germanic languages originate from words with a similar original meaning to that of 'science', namely as a general synonym of 'knowledge'. *Wissenschaft* is now similar in meaning to 'science' but with the important difference that it covers all the academic fields, including the humanities.

The terminology is not important. What is important is the existence of a community of knowledge disciplines, each of which searches in a systematic way for valid knowledge in its own subject area. Due to their different subject areas the knowledge disciplines differ widely in their methodologies

– from interpretations of ancient texts to calculations based on recordings of particle collisions in a synchrotron. Nevertheless, they are united by a set of common values, including the tenet that truth claims should be judged according to universal and impersonal criteria, as independently as possible of the value-based convictions of the individual scientist. Importantly, the knowledge disciplines are also connected through an informal but nevertheless well worked-out division of intellectual labour. The disciplines that are part of this community respect each other's competences. For instance, it is self-evident that biologists studying animal movement rely on the mechanical concepts and theories developed by physicists, that astronomers investigating the composition of interstellar matter rely on information from chemistry on the properties of molecules, and so on. This mutual reliance also applies across the supposed barrier between the 'sciences' (in the traditional, limited sense) and the humanities. An astronomer who wishes to understand ancient descriptions of celestial phenomena has to rely on philologists regarding issues of text interpretation, and similarly an archaeologist has to ask medical scientists for help in studies of disease risks in the ancient world.

Thus, the community of knowledge disciplines includes not only those usually called sciences, but also others that fall under the designation of *Wissenschaft*, including for instance literature and the history of arts. In its entirety the community covers a wide array of subject areas that can be summarised under five headings:

- nature (natural science);
- ourselves (psychology and medicine);
- our societies (social sciences);
- our own physical constructions (technology and engineering);
- our own mental constructions (linguistics, literature studies, mathematics and philosophy).

To simplify the nomenclature in the following, we will use the term 'science' in a broad sense that captures the disciplines denoted in German as *Wissenschaften*. Our topic is the position (if any) of risk analysis in this community of knowledge disciplines, not only among the disciplines conventionally covered by the English word 'science'.

THE NATURE OF SCIENTIFIC KNOWLEDGE

Many attempts have been made to specify the type of knowledge that is characteristic of science by means of specifying or delimiting the methods or methodologies that give rise to scientific knowledge. Probably the best known among these is Karl Popper's falsifiability criterion, according to which "statements or systems of statements, in order to be ranked as scientific, must be capable of conflicting with possible, or conceivable observations" (Popper 1962, p. 39). This and other such proposals are intended to be directly applicable to concrete issues of demarcation. For any given activity, such a criterion should be able to tell us whether or not it is scientific. However, all such criteria have severe problems. Most of them are suitable for some but not all of the disciplines of science, and they tend to exclude the science of previous centuries as unscientific, although it was the best of its day.

It should come as no surprise that definitions of science that require adherence to some specific scientific methodology run into difficulties. What unites the sciences, across disciplines and over time, is the basic commitment to finding the most reliable knowledge in various disciplinary areas. The term 'reliability' is used in the standard epistemological sense of being obtained in a truth-conducive way; for a detailed discussion of the concept, see e.g. Hudson (1994) and Lemos (2007). However, the precise means to achieve this differ among subject areas, and the chosen methods are also continuously in development. The major strength of science is its capability for self-improvement. Many of its most important self-improvements have been methodological, and these improvements have repeatedly been so thorough as to change not only the detailed methods but also high-level general methodological approaches, including principles for hypothesis testing, the acceptability of different types of explanations, and general experimental procedures such as randomisation and blinding. Therefore, a methods-based delimitation of science can only have temporary validity.

For a principled discussion such as ours, a definition of science is needed that is fully general and therefore applicable to science at different stages in its historical development. Consequently, such a definition cannot by itself determine in each particular case what is and what is not science. Since the purpose of science is to provide us with reliable knowledge, we can specify science as follows, in line with the definition stated at the beginning of Section 3.2:

> Science (in the broad sense) is the practice that provides us with the most reliable (i.e. epistemically most warranted) statements that can be made, at the time being, on subject matter covered by the community of knowledge disciplines, i.e. on nature, ourselves as human beings, our societies, our physical constructions, and our thought constructions.
>
> (Hansson 2013)

It should be emphasised that although the concept of reliability, in the sense just explained, can do much of the work in demarcating the outer boundaries of science, for the internal issue of determining scientific quality it has to be supplemented with additional criteria such as explanatory power and scientific fruitfulness (capability of contributing to the further development of scientific theories and scientific knowledge in general).

NORMS AND VALUES IN SCIENCE

Scientific practice is characterised by a set of norms and values that is remarkably similar across the disciplines. Historians, mathematicians and biologists alike expect their colleagues to be open to criticism, to disclose information that speaks against their own hypotheses, and to attempt to get every detail right. An influential attempt to summarise these standards was made by Robert K. Merton (1973). According to Merton, science is characterised by an ethos or spirit that can be summarised as four sets of institutional imperatives.

The first of these, *universalism*, asserts that whatever their origins, truth claims should be subjected to pre-established, impersonal criteria. This implies that the acceptance or rejection of claims should not depend on the personal or social qualities of their proponents. The second imperative, *communality*, says that the substantive findings of science are the products of social collaboration and therefore belong to the community, rather than being owned by individuals or groups. His third imperative, *disinterestedness*, imposes a pattern of institutional control that is intended to curb the effects of personal or ideological motives that individual scientists may have. The fourth imperative, *organised scepticism*, implies that science allows detached scrutiny of beliefs that are dearly held by other institutions. This is what sometimes brings science into conflict with religions and other ideologies.

In popular discussions, science is often described as being ideally 'value-free'. That is impossible; values such as those associated with Merton's four imperatives are necessary as guidelines in scientific activities. However, there is an important kernel of truth in the ideal of a value-free science. Although science neither can nor should be free of all values, there are many types of values that we require scientists to be as little influenced by as possible in their scientific work. In particular, we expect their factual statements to be as unaffected as possible by their religious, political or other social convictions.

There are two categories of values (and norms) that have a claim to be accepted in science and integrated in the scientific process. The first of these are what philosopher Carl Hempel termed the 'epistemic' values (Hansson and Aven 2014). These are values that support the scientific process itself: the values of truth and error-avoidance, the values of simplicity and explanatory power in hypotheses and theories. The presence of such values in science is generally recognised by philosophers of science.

The second category is much less commonly discussed or even recognised: non-controversial social and ethical values, i.e. non-epistemic values that are shared by virtually everyone or by everyone who takes part in a particular discourse. The presence of non-controversial values in science is often overlooked, since we tend not to distinguish between a value-free statement and one that is free of controversial values. Medical science provides good examples of this. When discussing antibiotics, we take for granted that it is better if patients survive from an infection than if it causes their death. There is no need to interrupt a medical discussion in order to point out that a statement that one antibiotic is better than another depends on this value assumption. Similarly, in economics it is usually taken for granted that it is better if we all become richer. Economists have sometimes lost sight of the fact that this is a value judgement. Obviously, a value that is uncontroversial in some circles may be controversial in others. This is one of the reasons why values believed to be uncontroversial should be made explicit, and not treated as non-values. Nevertheless, the incorporation of uncontroversial values such as the basic precepts of medical ethics will have to be recognised as reasonable in applied science, provided that these values are not swept under the rug but instead are openly discussed and taught and put to question whenever they become less uncontroversial than they were thought to be.

3.2.3.2 Risk analysis and science

We now proceed to characterise the field of risk analysis in terms of its subject matter, its relationship to risk management and the major types of studies that are performed within the field.

THE SUBJECT MATTER OF RISK ANALYSIS

Humans have worried about risks, and tried to manage them, since prehistoric times. Here our focus is on the modern field of risk analysis, a research area that was established in the 1970s. From its beginnings it was unusually interdisciplinary, drawing on competence in areas such as toxicology, epidemiology, radiation biology, and nuclear engineering. Today risk analysis has many other applications, including air pollution, climate change, preservation of cultural heritage, traffic safety, criminality and terrorism, and efforts to detect asteroids or comets that could strike Earth, to mention just a few examples. Many, if not most, scientific disciplines provide students of risk with specialised competence in the study of one or other type of risk – medical specialities are needed in the study of risks from diseases, biological specialities in studies of environmental risks, engineering specialities in studies of technological failures, etc. In addition, quite a few disciplines have supplied overarching approaches to risk. Statistics, epidemiology, economics, psychology, anthropology, sociology, and more recently philosophy are among the disciplines that have developed general approaches to risk, intended to be applicable to risks of different kinds.

RISK ANALYSIS AND THE MANAGEMENT OF RISK

As an example, consider a proposal to start employing a new vaccine for human use. Analyses and tests are carried out to determine its effects on humans in both the short and the long run. Let us assume that the results at a given stage all indicate that the vaccine has the desired preventive effect but no significant negative health effects. The authorities conclude that it is safe and introduce it for use. We can illustrate the most basic conception of this science-policy process as shown in Figure 3.4.

The data and information provided by the various analyses and tests supply the premises for constructing a knowledge base which is the collection of all 'truths' (legitimate truth claims) and beliefs that the majority of scientists in the area take as given in further research in the field. In the vaccine case, the knowledge base was considered to be so clear that a decision on use was justified. In a decision like this the role of the knowledge base is to support a decision, and then a judgement is made that takes us beyond the purely scientific sphere. The evidence and the knowledge base are supposed to be free of non-epistemic values ('value-free' in the common but misleading jargon). Such values are presumed to be added only in the third stage. Concluding that a vaccine is safe enough is a judgement based both on science and on values. This is necessarily so, since the question "How safe is safe enough?" cannot be answered by purely scientific means.

However, in practice the process is usually much more convoluted than this diagram indicates. The interpretation of the knowledge base is often quite complicated since it has to be performed against the background of general scientific knowledge. We may have tested the vaccine extensively and studied its mechanism in great detail, but there is no way to exclude very rare effects or effects that will only materialise twenty years into the future. Although the decision to disregard such possibilities is far from value-free, it cannot in practice be made by laypeople, since it requires a deep understanding of the available evidence seen in relation to our general biological knowledge. Based on this we need to add a risk evaluation step into the process, as shown in Figure 3.5. This is a step where the knowledge base is evaluated and a summary judgement is reached on the risk and uncertainties involved in the case under investigation. This evaluation has

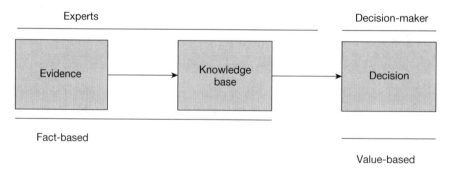

Figure 3.4 A first approximation of the information flow in the process in which science is used as a basis for decision-making about risks (based on Hansson and Aven 2014)

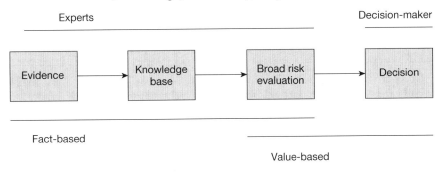

Figure 3.5 A second approximation of the information flow in the process in which science is used as a basis for decision-making about risks (based on Hansson and Aven 2014)

to take the values of the decision-makers into account (in this case, how safe they require the vaccine to be), and a careful distinction has to be made between the scientific burden of proof – the amount of evidence required to treat an assertion as part of current scientific knowledge – and the practical burden of proof in a particular decision. However, the evaluation is so intertwined with scientific issues that it nevertheless has to be performed by scientific experts. Many of the risk assessment reports emanating from various scientific and technical committees perform this function. These committees regularly operate in a 'no man's land' between science and policy, and it is no surprise that they often find themselves criticised on value-based grounds.

But it does not end there. In social practice, risk issues are often connected with – or rather, they are parts of – other issues. Decision-making on vaccination has to be based not only on information about risks but also on the costs of a vaccination campaign, its chances of achieving sufficient coverage, the severity of the disease in different strata of the population, alternative uses of the same resources, etc. Similarly, decision-making on traffic safety has to be integrated with decision-making on traffic planning as a whole, including issues such as travel time, accessibility, environmental impact, costs, etc. Although it is common in risk analysis to treat risk issues as separate from other issues in the same social sector, actual decision-making usually has to cover a larger group of concerns than those that are treated in risk analysis. Therefore, even after the risk evaluation, decision-makers need to combine the risk information they have received with information from other sources and on other topics. In Figure 3.6 we refer to this as the decision-maker's review and judgement. It clearly goes beyond the scientific field and will cover value-based considerations of different types. It may also include policy-related considerations on risk and safety that were not covered in the expert review. Just like the expert review, however, it is based on a combination of factual and value-based considerations.

It is stressed that Figure 3.6 is a model, a simplification of the real world, and this is even more true of Figures 3.4 and 3.5. The arrows represent the major flows of decision-guiding information. Obviously there are also other communications taking place, for instance questions from decision-makers to scientists and feedback from risk assessors to scientists producing risk-relevant data. We have chosen not to complicate the model by including these types of communication.

TWO MAJOR TYPES OF RISK STUDIES

The content of risk analysis can be inventoried by reviewing different types of papers published in relevant scientific journals. Based on the discussion by Aven and Zio (2014) on the foundations of risk assessment and risk management, we can define risk analysis as comprising: (i) knowledge about risk-related phenomena, processes, events, etc.; and (ii) concepts, theories, frameworks, approaches, principles, methods and models to understand, assess, characterise, communicate and (in a wide sense) manage risk, in general and for specific applications. This provides us with a division of risk studies into two major categories. One of these consists of papers aimed at knowledge about risk-related phenomena, processes, events, etc. – for example, the consequences of using a specific drug in a medical context, the damage to the environment caused by an oil spill in a specific type of coastal area, etc. Such knowledge is usually obtained by combining insights from different disciplines, for example medicine and biology, with various formal modelling approaches, most commonly traditional statistical analysis involving hypothesis testing or Bayesian analysis. These studies are mostly concerned with issues arising in the processes represented by the two leftmost rectangles in Figure 3.6.

Another main category of papers is related to the development and analysis of new concepts, theories, frameworks, approaches, principles,

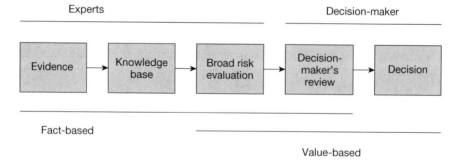

Figure 3.6 Our third and final approximation of the information flow in the process in which science is used as a basis for decision-making about risks (based on Hansson and Aven 2014)

methods and models for the understanding, assessment, characterisation, communication, management and governance of risk – in short, instruments for assessing and managing risk. For example, consider the development of an appropriate risk concept. During the last thirty years we have seen a number of suggestions on how to define and understand this concept, but the area is still struggling to establish a consensus. Another example is the development of suitable risk analysis methods to reflect human and organisational factors. In recent years several different research environments have presented and discussed new frameworks and methods for obtaining more realistic and detailed representations of such factors. These are just examples of the considerable amount of ongoing work, which aims to find suitable instruments for assessing and managing risk.

There are also other types of papers on risk published in scientific journals, but these two broad categories are in our view the most important ones. They provide the basis for the risk-decision process, as outlined under the heading 'risk analysis and the management of risk'.

3.2.3.3 Is risk analysis scientific?

We will now return to the three aspects of science outlined above, namely, its subject matter, its type of knowledge, and its values and norms, in order to answer the question of whether risk analysis is scientific.

ITS SUBJECT MATTER

Risk studies are involved in all the five main areas of science mentioned above. These include studies of nature (e.g. toxicology, volcanology and climatology), ourselves (risk perception), our societies (risk governance, economics), our physical constructions (technical risk analysis, safety engineering), and our mental constructions (risk conceptualisations, the ethics of risk). Although risk analysis comprises an unusually extensive research field, it would seem to be difficult to find any part of it that falls outside of science in the broad sense that we use the term here: namely, the community of knowledge disciplines that also includes the humanities.

THE NATURE OF ITS KNOWLEDGE

It is quite straightforward to characterise as scientific the first of the two types of risk studies that we referred to in Section 3.2.3.2. Finding out the toxicity of a chemical substance, the effects of carbon dioxide emissions on the climate or the behavioural effects of different person–machine interfaces are all activities that fall squarely under the designation of science – provided, of course, that the most reliable knowledge is consistently sought.

The models developed in the second type of risk studies are arguably somewhat more problematic in this respect. Of course, model building per se is almost ubiquitous in science, in particular in scientific theories. A model is a system that has been constructed to focus on selected important properties of the phenomena under study. The relations between different phenomena are often expressed in the form of mathematical equations, whose numerical variables correspond to different quantities which the model represents. Mathematical models have been indispensable since the nineteenth century in physics, chemistry and technology, and in economics since the middle of the twentieth century.

Models can be either deterministic or probabilistic. In risk analysis we model variabilities, and probabilistic models have turned out to be the most useful for the majority of purposes. Such probabilistic models contribute to our knowledge about risk-related phenomena, processes, events, etc. However, it must be understood that as with all scientific models, these models are based on idealisations (simplifications) in relation to the phenomena that are modelled. This means that some of the complexities of the study object are excluded in order to capture the relationships between what we consider to be its most important features. When we are dealing with complex social phenomena, as we typically are in issues of risk, the choices regarding what to include in a model and how to represent the included factors are always to some extent subjective. However, this is not an uncommon situation in science. Experience shows that the fastest road to scientific progress is often to pursue the best models that we presently have, even if they are known to be imperfect. In this way we can understand their imperfections more thoroughly and learn how to build better models.

For some of the variables that we wish to include in our models, meaningful probabilities may not be available. For instance, this often applies to the possibility that there may be unknown dangers associated with some new, uninvestigated phenomenon. A common approach in risk analysis is to probabilise even that which is difficult to probabilise, often with probability estimates that have been elicited from experts. Unfortunately, such estimates may be severely wrong in spite of the experts' best efforts. It is therefore essential to distinguish between such uncertain probability estimates and probabilities that have a safe epistemic grounding – for instance, in frequencies derived from relevant experience.

Virtually all probabilities (defined through probability models or as probability model parameters) that we deal with in risk analysis are to some extent uncertain – their 'true' values are not known with certainty. In some cases the uncertainty is small enough to be negligible. In other cases the uncertainty is non-negligible, but unimportant for most practical purposes. If we are told that the estimated probability of a severe but preventable accident is 0.02, this is usually sufficient for practical action. Information that the probability might possibly be as low as 0.005 or as high as 0.1 would in most cases make no real difference to the decision-maker. However, there are other cases when uncertainty about probabilities may have a practical impact. A decision-maker who accepts a probability of 10^{-6} of a fatal accident might have reacted differently if told that this was an uncertain estimate and that other equally competent experts have estimated the probability at 10^{-4}.

Unfortunately, the uncertainty associated with probability estimates has often been neglected. Once a probability estimate has been made, we have a strong tendency to treat it as reliable. This may have negative effects on decision-making, but the fault is not necessarily located in the probabilistic model and the analysis. Usually the crucial issue is how such models are interpreted. In communicating risk research it is a major challenge to distinguish adequately between the risk described through a model and the uncertainty inherent in that model itself.

A fundamental topic in risk analysis is the representation and treatment of uncertainty (i.e. epistemic uncertainty) about such probabilities and other unknown quantities. Different methods are available for dealing with these uncertainties. Following the traditional statistical approach, confidence intervals are used to describe uncertainties, but of course this measure only reflects statistical variation. Bayesian analysis is based on subjective probabilities expressing the assessor's degree of belief in the occurrence of the event. A wide variety of other methods and models have been developed, for instance making use of interval (imprecision) probabilities. See Appendix A and Aven *et al.* (2014a) for further details.

The representation of epistemic uncertainty is challenging. For example, it is not easy to include in a probability calculation a numerical value of the uncertainty representing the possibility that the calculation may itself be wrong. A probabilistic framework is also difficult to apply in the analysis of security issues where an adversary performing intentional actions is involved.

A main problem is the treatment of rare events, and in particular black swan events. For such events the traditional scientific perspective of risk analysis, attempting to accurately estimate risk and predict the events, fails. However, risk analysis can still provide informative risk and uncertainty descriptions, as argued in this book.

Risk analysts can conceptualise and partially characterise the risks associated with such events and develop methods for identifying and dealing with them. Since there are always limited resources available for managing these risks, such assessments are needed not least for priority-setting. The decision makers need to be informed about issues related to important precursors, the knowledge available and the uncertainties.

Thus, although risk analysis cannot produce accurate probability estimates of surprising and/or extremely rare events, useful science-based decision support can take other forms such as the following:

- Characterisations of the robustness of natural, technological, and social systems and their interactions.
- Characterisations of uncertainties, and of the robustness of different types of knowledge that are relevant for risk management, and of ways

in which some of these uncertainties can be reduced and the knowledge made more robust.

- Investigations aimed at uncovering specific weaknesses or lacunae in the knowledge on which risk management is based.
- Studies of successes and failures in previous responses to surprising and unforeseen events.
- Development, analysis, and measurement of "generic abilities" that are considered needed in the response to a wide range of surprising and unforeseen events.

Having defined science as the practice that provides us with the most reliable (i.e. epistemically the most warranted) statements that can be made on the issue, one might question what the most reliable statements on this issue are. Different schools and authors argue for completely different approaches and perspectives; see Appendix A and Aven *et al.* (2014a).

What we discuss here is the second type of study addressed above, linked to concepts, approaches and methods used in risk analysis (the instrumental part). The scientific literature consists of papers addressing the issue of how best to represent and treat epistemic uncertainties. Many suggestions have been put forward and arguments provided for their use, as mentioned above. We can consider all these contributions as inputs to the knowledge base (see Figure 3.6). However, they do not allow for very far-reaching conclusions, and therefore this knowledge base is not determinate enough to tell us unambiguously which approach to apply in practice. It is not at all clear which statements are the most epistemically warranted. The area is currently characterised by different perspectives and a lack of consensus. To make a decision on which approach to use in a specific case, there is a need for judgements that go beyond the traditional scientific sphere; we are in stage four of Figure 3.6. If the scientific community had been able to establish a broad, science-based consensus on an approach to represent and treat epistemic uncertainties, then the decision-maker's review and judgement would have been a much easier task.

Scientific papers on the representation and management of uncertainty provide knowledge mainly through argumentation. We gain knowledge about a new way of representing the uncertainties, how it relates to other approaches, the pros and cons of using this approach, etc. This clearly adds new insights to the knowledge base. A contribution may also provide an argument for this new approach to replace others, attempting to show that it is preferable. Then normative aspects have entered the scene, but not all of this argumentation should be considered as non-scientific (at the stage of the decision-maker's review and judgement). Provided that the analysis is solid (precision in concepts, consistency ensured, etc.), the analysis adds valuable insights to the knowledge base by explicating the arguments leading to the conclusions. There are value judgements in this process, but the argumentation should enrich the knowledge base since it links the premises of the new approach to the decision-making context. Although other weights could be given to the various arguments, and hence other conclusions could be drawn, there can still be a strong scientific knowledge element in the argumentation.

We have used epistemic uncertainty representations as an example, but the discussion is relevant for any instrument (concept, approach, method) to be used for assessing and managing risk. The scientific literature covers a lot of work related to finding suitable instruments for assessing and managing risk. As argued above, this work is on the borderline between science and value judgements, with one main part in the knowledge base, and one part in managerial review and judgement.

ITS NORMS AND VALUES

Non-controversial values can legitimately be incorporated into applied science. For instance, this applies to basic values in clinical medicine concerning the physician's commitment to further the patient's health. Such non-controversial values can also be found in risk studies. It is uncontroversial that a reduction in the frequency or severity of accidents is per se desirable, that a reduced cost of safety measures is also desirable, that decreased environmental damage is desirable, etc. However, contrary to clinical medicine, risk analysis is also confounded by value issues that are far from non-controversial. This applies in particular to cases in which risk reductions are associated with substantial (monetary or non-monetary) costs. How much should we be willing to pay to avoid a traffic fatality? How large a risk should rescue workers accept in order to save an avalanche victim in a highly dangerous area? How should we weigh the ecological value of saving the Indian tiger against the dangers to humans from tiger attacks? Issues like this are closely connected with risk analysis, and risk analysts have often tried to address them. Obviously, knowledge about facts, probabilities, and uncertainties are not sufficient to provide answers.

In the history of risk analysis, it has repeatedly been proposed that some specific way to answer such questions is the only rational way to do it. Currently the most popular such proposal is to let the outcome of cost–benefit analysis have the final say on what decision to make. Although cost–benefit analysis undoubtedly provides highly valuable information for decision-makers, its use as the exclusive and final decision criterion is controversial, and this – at least in part – is for good reasons. As already noted, risk decisions tend to be integrated in more general decision complexes. It is not always an advantage to separate out risk issues and treat them as isolated problems. Furthermore, the principle according to which one should pay the same price for a reduction in the expected number of saved lives in all social sectors cannot be implemented without a centralisation or co-ordination of economic decision-making, which may have considerable drawbacks from the viewpoint of efficiency.

In addition, the use of expected values in cost–benefit analysis to a large extent ignores uncertainties (see Section 5.1.1). It would be unwise to assume that this or any other method to weigh different risks and benefits against each other represents the only rational way in which decisions under risk and uncertainty can be made. Instead, the value component of risk decisions should be displayed as clearly as possible, and risk scientists should limit their efforts to the first three rectangles in Figure 3.6, leaving the last two to decision-makers. This is admittedly a difficult line to draw, not least since some value issues have to be dealt with by risk scientists (the third rectangle). Nevertheless, drawing this boundary is essential for the public credibility of risk analysis.

3.2.3.4 Conclusion

We conclude that risk analysis is scientific, when understood as consisting primarily of: (i) knowledge about risk-related phenomena, processes, events, etc.; and (ii) concepts, theories, frameworks, approaches, principles, methods and models to understand, assess, characterise, communicate and manage risk, in general and for specific applications (the instrumental part). However, risk analysis still has a way to go to clarify its scientific foundation, as illustrated by the above discussion of the treatment of uncertainty. The model represented in Figure 3.6 is developed to provide guidance for the process in which science is used as a base for decision-making on risk. Through the model, the spheres of the experts and the fact-based areas are delimited, as well as those of the decision-makers and the value-based areas. The above analysis has underlined the importance of the third and fourth elements in Figure 3.6 (the broad risk evaluation and the managerial review and judgement). However, there is a further need to clarify the content of these stages. Considerable scientific work in risk analysis has been devoted to identifying and developing suitable instruments and finding those that are best in some sense. This work adds insight to the knowledge base by balanced arguments and reasoning, but we quickly move into the normative sphere when concluding what we ought to do – for instance, that we should use a particular concept or method that we have found to be more suitable than others. We have referred to various scientific and technical committees that play an important role in stage three, the risk evaluation. Such committees have to a large extent been limited to the first part of risk analysis, making judgements about risk-related phenomena, processes, events, etc. However, we also need to have arenas where risk scientists evaluate the instrument evidence base. There are a number of groups working on such issues, but we are not aware of broad authoritative scientific committees acting today with such a purpose. Although history has shown that it is difficult to obtain consensus in broad committees on instruments (Thompson *et al.* 2005), the field needs such arenas to stimulate deliberations and debate.

3.3 Uncertainty

When discussing uncertainties in a risk context we need to clarify:

(i)   What are uncertain? (Sections 3.3.1 to 3.3.3).
(ii) Who is uncertain? (Section 3.3.4).
(iii) How should we represent the uncertainties? (Appendix A).

As mentioned in the introduction to this chapter, we distinguish between three main categories of (i):

- uncertain quantities (including the occurrence or non-occurrence of events) (Section 3.3.1);
- the future (Section 3.3.2);
- phenomena (Section 3.3.3).

The concept of model uncertainty is discussed in Sections 3.3.1 and 3.3.3. When it comes to (ii), we question whether it is the decision-maker, the analysts or some other experts used in the assessment who are uncertain. In order to obtain a clear understanding of the risk and uncertainties and to communicate relevant results, it is essential to be precise on this issue.

To express the uncertainties an adequate representation is required, and probability is the natural choice since it meets the basic requirements for such a representation (Bedford and Cooke 2001, p. 20):

- *Axioms*: Specifying the formal properties of the uncertainty representation.
- *Interpretations*: Connecting the primitive terms in the axioms with observable phenomena.
- *Measurement procedures*: In conjunction with supplementary assumptions, providing practical methods for interpreting the axiom system.

However, in recent years we have seen a growing interest in alternative representations, based on interval probabilities, possibility theory and evidence theory. Although this book does not discuss these alternative approaches in detail, some comments are made in this chapter (in Section 3.3.1; see also Appendix A).

3.3.1 Uncertainty of a quantity

We face the issue of characterising uncertainties about an unknown quantity, and in particular using probabilities to express this uncertainty (i.e. epistemic uncertainties) in a risk assessment context. A basic introduction to this topic is given in Section 2.2.

To express uncertainties about an unknown quantity, the common approach is to use a subjective (knowledge-based, judgemental) probability, expressing the assessor's uncertainty (degree of belief) about the occurrence of event A. We denote this probability P, or P(A|K) to show that this probability is conditional on some background knowledge, K. A common interpretation is the uncertainty standard, the probability P(A) = 0.1 (say), means that the assessor compares his/her uncertainty (degree of belief) about the occurrence of the event A with the standard of drawing at random a specific ball from an urn that contains ten balls.

Let us return to the offshore installation example (Section 1.3) and assume an assigned probability of 1/10000 that an accident will occur causing at least one hundred fatalities. If subjective probabilities, according to the uncertainty standard interpretation, had been the basis of the analysis, the analysts would explain the meaning of this probability in this way: based on the analysts' knowledge (assumptions, models, data), the assigners' uncertainty related to the occurrence of the event and the analysts' degree of belief that the event will occur are the same as drawing a specific ball from an urn containing ten thousand balls. The interpretation would preferably be explained once at the beginning of the presentation of the results, as a part of the overall introduction to the scientific framework adopted. Following this interpretation, the results of the assessment cover the analysts' judgements concerning their uncertainty and degrees of belief for the events of interest, but the analysts' value judgements concerning different outcomes are not included. The analysts communicate their pure uncertainty assessments.

The assignment of subjective probabilities following this interpretation can be carried out using different approaches, as discussed for example by Lindley (2006) and Aven (2008, 2012a). Direct probability assignment by the analysts can be used based on all the sources of information available, formal expert judgements, modelling and Bayesian analysis using probability models, and specification of prior probability distributions for unknown parameters; see, for example, Aven (2008, 2012a). These references provide a number of examples for how to derive the probabilities for real-life examples. In his book *Understanding Uncertainty* (2006), Lindley provides a thorough discussion of the concepts introduced, including the urn set-up, but of course this set-up need not and should not be referred to in all assignments of probability. Having carefully explained what a probability of (say) 0.2 means, the many probabilities produced by the risk analysis can be reported without explicitly mentioning this standard.

To adequately use the results of a risk analysis, a clear understanding of how to interpret the probabilities is critical. The offshore installation example in Section 1.3 illustrates this. In this example the management, i.e. the decision-maker, was informed by the probability numbers, but the lack of clarity about what these numbers expressed made them question how to use the results of the analyses. The interpretation discussion was not an academic one; it was considered a critical issue to be solved in order for the management to be able to properly understand what the risk analysis was actually communicating. Without clarity regarding the meaning of the probabilities reported, the managers did not know how to use them in the decision-making process.

This case is an example from the oil and gas industry. However, the issue is the same in other industries, for instance the nuclear industry. Probabilities are used to represent and describe uncertainties in quantitative risk analyses (probabilistic risk analyses), but when it comes to subjective probabilities reflecting epistemic uncertainties, clear interpretations are often lacking. The analyses are intended to inform the decision-makers, as in the platform case, but how can that be properly achieved when it is not explicitly formulated what 0.3 means compared to (say) 0.2? A risk-reducing measure may be considered which reduces the assigned probability from 0.3 to 0.2, but without interpretation the importance of this risk reduction cannot be fully understood. An analogous argument would apply when it comes to checking whether intolerability limits have been exceeded or not: if the assigned probability is below the limit, the decision-maker's actions are dependent on what the probabilities are actually saying. Is the probability simply a subjective judgement made by the analysis team based on their background knowledge, or is the probability trying to represent the data and knowledge available in a more 'objective' way? Clearly, these two types of interpretations could lead to different conclusions in practice. In the latter case, a probability number below the intolerability level has a much stronger decision strength than in the former case. Thus, for proper use of the risk analysis a clarification is required. If we are in fact using judgemental probabilities (the former case), we have to communicate this in a way that allows the managers and the decision-makers to adequately reflect this in the decision-making process. If the latter case can be applied, there is often a short way to risk-based decision-making (in contrast to risk-informed decision-making) – as, for instance, the intolerability example demonstrates: the probability number is below the limit, and as the number is to a large extent 'objective', the risk is considered tolerable. We refer to this issue in Section 5.1.2.

Using knowledge-based (subjective, judgmental) probabilities to quantify uncertainties means that the analysts must express their degree of belief about unknown quantities using probability distributions. For example, if it is known that a physical quantity has a value between 0 and 1, the analysts may specify a uniform distribution of [0,1] to express their uncertainties. The analysts have then assigned the same probability (1/2) for the quantity to be in the interval [0, ½] as in the interval [½, 1]. Following a probability-based approach, such assignments are required. However, this perspective can be challenged: the assessments are based on unjustified assumptions. Instead of specifying one distribution we could alternatively specify upper and lower probability distributions, reflecting that the information does not allow us to be more precise. We are led to interval probabilities. In the literature there is an ongoing discussion about the suitability of such alternative approaches to exact probabilities; see the discussion in Aven (2010b), North (2010) and Dubois (2010), as well as Aven *et al.* (2014a).

A key issue in the discussion is whether we seek:

(1) to obtain an objective (inter-subjective) description of the unknown quantities; or
(2) to obtain a judgement about the unknown quantities from a qualified group of people (the analysts).

As argued in Aven (2010b), we need the subjectivist probability approach to inform decision-makers, reflecting the judgements of qualified analysts and experts. However, we also need in some way to address the fact that decision-makers and other stakeholders may find this resulting risk picture insufficiently informative. If we seek to obtain a more 'inter-subjective' knowledge description of the unknown quantities studied, the use of imprecise probabilities could add information to the precise probability approach.

This is an area that needs further development. There is a huge mathematically-oriented body of literature on interval (imprecise) probabilities (for example, supported by possibility theory and evidence theory), but relatively little focus has been placed on discussing how these intervals are to be used in practical contexts. For example, what type of assumptions are we allowing when we are expressing interval probabilities? Fewer assumptions mean a higher level of inter-subjectivity, but broader and rather non-informative intervals. See discussion in Aven (2010b) and Dubois (2010).

The probabilities reported in the offshore installation example (see Section 1.3) were precise numbers, but it is possible in theory to think about an analysis framework which also allows for imprecise probabilities. However, how to carry out such an analysis is not straightforward given the hundreds of probability assessments to quantify and integrate; see also the discussion in Aven and Zio (2011). In any case, such an interval analysis should be a supplement to the precise assignment processes, not a replacement.

We refer to Aven *et al.* (2014a) for an in-depth analysis of such probabilities and their link to possibility theory and evidence theory. The present book is based on the conviction that to describe risk, the probabilistic approach is useful but it needs to be supplemented by alternative approaches. We will return to this idea in Chapter 4.

Next we will discuss uncertainties in relation to models, the parameters of the models and the structures. To express the uncertainties we often use models, such as event trees and fault trees, and in particular probability models (i.e. models that are based on frequentist probabilities; see Section 3.1 and Appendix A.1.2). By assessing uncertainties of the parameters of the models, uncertainties are described for the quantities of interest as explained in the following.

PARAMETERS AND MODELS

Consider a quantity Z whose true value is unknown. As an example, Z could be the actual number of fatalities due to a potential outbreak of a new virus. The actual value of Z cannot be known until after an outbreak. To predict the future value of Z, a model $G(X)$ is developed. Both X and Z may be vectors. A simple model would be $G(X) = G(X_1, X_2) = X_1 X_2$, where X_1 is the fatality rate and X_2 is the number of exposed people (say the number of citizens in a country). The predictions made by the model $G(X)$, then, depend on the structure G and the parameters X_1 and X_2.

Given the model G, a probability distribution can be computed for Z using standard probability calculus:

$$P(G(X) \le z) = \int P(G(x) \le z | X = x) \, dH(x) = \int_{\{x:\, G(x) \le z\}} dH(x),$$

where H is the probability distribution of X.

Define:

> *Model error*: The difference, $\Delta G(X)$, between the model prediction $G(X)$ and the true future value Z, i.e. $\Delta_G(X) = G(X) - Z$.

> *Model output uncertainty*: Uncertainty about the magnitude of the model error.

Note that according to this definition, model error and model output uncertainty are different but connected concepts. Model output uncertainty is actually *epistemic* uncertainty about the model error, and hence it may in theory be assessed using a suitable tool for measuring this type of uncertainty, such as subjective probabilities and interval probabilities. The above discussion about uncertainties regarding a quantity applies; see Bjerga *et al.* (2014).

The model output uncertainty results from two components:

> *Structural model uncertainty*: The conditional uncertainty associated with the model error $\Delta_G(X)$, given the true value X_{True} (i.e. $\Delta_G(X_{True})$).

> *Input quantity (parameter) uncertainty*: The uncertainty associated with the true value of the input quantity X.

The structural model uncertainty expresses the epistemic uncertainty under the condition that the input parameters are known (the true values). In other words, the structural model uncertainty expresses uncertainty about the model error, when we can ignore uncertainty about the parameters X, and then relates to the model structure G itself. Typically this uncertainty is associated with assumptions and suppositions, approximations and simplifications made in the modelling. Input quantity (parameter) uncertainty, on the other hand, reflects epistemic uncertainties about the 'true' value of X. In the context of probability models, the meaning of the 'true' value can be explained as follows.

A probability model is based on a set-up which is thought-constructed and refers to an infinite number of systems similar to the one under study. The meaning of 'similar' has to be clarified – in a Bayesian context it means that the different system quantities can be judged to be exchangeable (a sequence of random quantities is exchangeable if their joint probability distributions are independent of the order of the quantities in the sequence), whereas in a traditional statistical context 'similar' means that the quantities are independent and identically distributed. In this modelling set-up, one can refer to a 'true' distribution F describing the variation in the quantities studied, for example the number of events occurring, in the infinite population of similar systems. We write 'true' in quotes because its meaning exists only within this thought-constructed set-up. As a model of this 'true' distribution we introduce the probability model, for example a Poisson model. By extension of this reasoning, we can talk of a 'true' value of a parameter, (say) λ in the Poisson case, to be interpreted as the average number of events in the infinite population of similar systems. Both F and λ are unknown and must be estimated. In a Bayesian analysis, the focus is on the epistemic uncertainties about this 'true' value of λ (expressed as prior and posterior distributions). In a traditional statistical analysis, one tries to estimate the 'true' value of λ and give a confidence interval for it (for the sake of simplicity, it is common to say that one estimates and gives confidence intervals for λ). Normally, less attention is paid to the epistemic uncertainties related to the probability model itself, but from the above reasoning it appears that it makes sense to address this in relation to the deviation between the 'true' F and Poisson distributions. For readability we will in the following avoid writing 'true' in quotes.

Sources of structural model uncertainty stem from actual 'gaps' in knowledge, which can take the form of poor understanding of phenomena that are known to occur in the system as well as complete ignorance of other phenomena. This type of uncertainty can lead to 'erroneous' assumptions regarding the model structure. Other sources of structural model uncertainty stem from approximations and simplifications introduced in order to translate the conceptual models into tractable mathematical expressions.

To use a model, the model uncertainty must be considered acceptable. Uncertainty analysis is a tool to accredit a model so as to ensure a certain quality and possible certification. In the accreditation process, the understanding of the influence of uncertainties on the results of the analysis is of importance in order to adequately guide the uncertainty reductions. If the model considered cannot be accredited, remodelling is required.

In a case where experimental data are available for comparing G(X) and Z, a wide range of statistical methods exist that can be used for validation in order to accredit a model. These methods include both traditional statistical analysis and Bayesian procedures; see e.g. Kennedy and O'Hagan (2001), Meeker and Escobar (1998) and Zio (2006). Model validation is often linked to model verification (and is often referred to as Verification and Validation, or simply V&V), which is commonly understood as the process of comparing the model with specified requirements (Knupp 2002, McFarland 2008, Oberkampf and Trucano 2002). The verification part is obviously important in many contexts to produce a model that meets the specifications.

In some situations no experimental data exist at the time of the assessment, and this leads us away from classical statistical tools for validation and subsequent accreditation of the model. Instead, validation transforms into utilising expert/analyst argumentation based on established scientific theories and specific knowledge about the system which the model is intended to describe. See discussion in Aven and Zio (2013).

An important observation is that no restrictions pertain to utilising a pure probability-based approach. The above understanding of the model uncertainties allows for both probabilistic and non-probabilistic approaches and thereby injects flexibility into the uncertainty analysis, giving the opportunity to choose the approach that is judged to best represent/express the uncertainties in the context of the specific phenomena and the surroundings examined.

POISSON EXAMPLE

The case study pertains to a risk assessment context for a specific activity, where the modelling of the uncertain future occurrences of a type of undesirable event is described by a Poisson model. Let $N(t)$ be the number of events occurring in the time interval $[0,t]$. It is assumed that the stochastic process N is a homogeneous Poisson process with occurrence rate λ. Hence $N(t)$ has a Poisson distribution with expected value λt, i.e. $P(N(t) = n \mid \lambda, t) = p(n \mid \lambda, t) = (\lambda t)^n e^{-\lambda t}/n!$, $n = 0, 1, 2, \ldots$ We interpret λ as the expected number of events occurring per unit of time.

Furthermore, let $p_0(n \mid t)$ be the true distribution of the number of events in $[0,t]$, obtained by considering an infinite number of activities similar to the one considered. The average number of events per unit of time is defined as λ_0. The Poisson distribution $p(n \mid \lambda, t)$ is a model of this true distribution, and λ_0 is the (unknown) true value of the model parameter λ. The distributions p_0 and p represent the true variation in the number of events occurring in such intervals and the variation as modelled, respectively.

In the case study, the objective of the risk assessment is to verify that the 0.95 quantile, n_{95}, of $p_0(n \mid t_0)$ is in compliance with a regulatory threshold value n_M, where t_0 is a fixed point in time.

In line with the above definitions, we identify n_{95} as the quantity of interest, Z, λ as the parameter X, and the model representing Z, $G(\lambda)$, as the 0.95 quantile of the Poisson distribution, which we refer to as $n_{95}(\lambda)$. The model error can thus be written $\Delta_G(\lambda) = G(\lambda) - n_{95} = n_{95}(\lambda) - n_{95}$. The structural model uncertainty relates to uncertainty about the value of $\Delta_G(\lambda_0) = G(\lambda_0) - n_{95} = n_{95}(\lambda_0) - n_{95}$ and the parameter uncertainty to the true value of λ, i.e. λ_0.

As a concrete example of this setting, we can consider potential releases from a commercial pilot facility/system handling crude oil with new technology, with a project period of five years. This system is operating in a seasonal market following economic cycles and variations in demand. The system is in the planning phase, and compliance with regulatory frameworks must be demonstrated prior to construction. Concerning the environmental risk and potential releases, the authorities acknowledge that releases could occur due to the novel technology and the limited operational experience. The authorities have specified an acceptance level of five releases during one month in order to license construction and continuous operation; the system must be demonstrated to be capable of meeting this level with a probability of 0.95.

For the model of the number of releases, a homogeneous Poisson process is initially found to be representative. The parameter λ representing the average number of releases is estimated to be 1.75 per month, based on an analysis of the technical solutions at the facility. The 0.95 quantile is calculated to $n_{95}(1.75) \approx 4$, and it is concluded that the requirement from the authorities is met (the probability of having more than five releases is calculated to be approximately 0.01).

But what about uncertainties, and the model uncertainties in particular? It is possible to use knowledge-based probabilities, but these are difficult to assign: the support for the numbers is hard to justify (Bjerga *et al.* 2014). Instead, a qualitative approach is recommended as illustrated in Section 4.2, see also Bjerga *et al.* (2014).

3.3.2 Uncertainty about the future

We consider an activity in the future, real or thought-constructed, and focus on the consequences C in the same way as we conceptualise risk by (C,U). Hence, uncertainty means not knowing what C will be, and to describe the uncertainties we are led to (Q,K) as for risk, seen in relation to the specific consequences C'.

Thus C is the true, actual consequences/outcomes of the activity. If the consequences are limited to C_1, for example an increase in sea level when discussing climate change, C = C_1, but C does not need to be described by a set of quantities as in C'. Consider the climate change example, and let us say that we have a hundred-year horizon. The true climate changes and the true C may not necessarily be well characterised by a set of quantities C' defined today. If we wish to formulate some specific consequences characterising the climate change, any choice of C' could easily lead to a poor characterisation of the true consequences. Another example may illustrate this better.

You are playing a game where you do not know the type of prize. Clearly any attempt to specify the outcomes could be extremely poor compared to the real outcomes, unless some general formulations are linked to attributes like human health, individual livelihoods, environmental conditions, economic aspects, etc. (see, for example, Aven and Renn 2010, p. 104).

However, for any uncertainty description we need specifications C', and we would normally seek to use some type of severity scale. As an example, return to climate change. Think about a quantity C_1 which equals the increase in sea level (suitably defined) over the next twenty years. Here we have a scale of severity defined, where a change of 1 metre could be said to be high and 0.1 low. Mathematically, we can define a set Ω of outcomes (outcome space) where the elements of Ω can be ordered according to a severity scale. In the sea level example $\Omega = [0,\infty)$, where Ω represents the set of increases in sea level, and if x and y are elements of Ω, x is less severe than y if x < y. Using probabilistic terminology, C_1 can be viewed as a random variable or a random quantity with values in Ω. When the activity is realised, C_1 takes one value in Ω. Before realisation, the space Ω expresses all possible outcomes that C_1 can take. In practice there could typically be many dimensions of C, C_1, C_2, ... with spaces Ω_1, Ω_2, ..., respectively. A severity scale can also be defined on the total integrated space, but it would in general be subject to more discussion – what is the appropriate weighting of the different components? This requires consideration of conflicting objectives and preferences. The severity scale for Ω_i would in general be more objective (inter-subjective) than the scale for the integrated space.

The specific consequences C' can also capture unknown variation metrics, such as frequentist probabilities (chances), parameters of probability models and the probability models themselves. Note that the C's need not be actually observed. They could be thought-constructed quantities trying to capture key features of C. In cases where we focus on probability models and related parameters, it is meaningful and informative in relation to C to look at the variation produced by generating a large number of similar units, for example the number of events in specific periods of time. If the situation considered is unique, probability models cannot be justified, and this type of C' is not introduced.

The choice of attributes to consider is not straightforward, as discussed by Renn and Klinke (2002). For many situations it is not enough just to focus on dimensions like the number of fatalities and economic loss. Equally important are aspects such as:

- *ubiquity* – which describes the geographical dispersion of potential damage;

- *persistency* – which describes the temporal extension of the potential damage;
- *reversibility* – which describes the possibility to restore the situation to the state before damage occurred;
- *delay effect* – which describes the period of latency between the initial event and the actual impact of damage;
- *violation of equity* – which describes the discrepancy between those who enjoy the benefits and those who bear the risk;
- *potential of mobilisation* – which is to be understood as violation of individual, social and cultural interests and values generating social conflicts and psychological reactions by individuals and groups on whom the risk consequences are inflicted; the potential of mobilisation could also result from perceived inequities in the distribution of risk and benefits.

Renn and Klinke's work was based on a classification scheme established as a part of a project conducted on behalf of the German government's Advisory Council on Global Change. Suitable C values can be defined based on these aspects, for example the time until at least x people die.

Next we will discuss the description of uncertainties, with reference to the previous section. Here we will stress that care has to be taken when comparing situations of large (small) uncertainties with situations of large (small) probability. Say that the measure Q = Probability. Then

Description of uncertainties = (P,K).

If K is ignored one is easily led into making interpretational mistakes. Some examples illustrate this.

Consider an unknown quantity X, which can take a value in the set {0,1,2,3}, with three associated cases:

(1) An aleatory uncertainty distribution P_f for X is known, expressing that $P_f(X = x) = ¼$, x = 0,1,2,3. This means that the X will take the value x in about 25% of the cases when performing repeated similar experiments.
(2) A knowledge-based probability distribution P for X is assigned, such that $P(X = x) = ¼$, x = 0,1,2,3. This means that the assigner's uncertainty and degree of belief is the same as the probability of drawing one particular ball out of an urn containing four balls.
(3) A knowledge-based probability distribution P for X is assigned, such that $P(X = 0) = 0.01$, $P(X = 1) = 0.04$, $P(X = 3) = 0.10$ and $P(X = 4) = 0.85$.

Where are the largest/smallest uncertainties? Where are the highest/lowest probabilities? If we compare the last two cases, it is clear that the uncertainties are largest in case (2). The assigner has a strong belief in a large outcome in case (3). In information analysis, the Shannon entropy H is often used as a measure of the uncertainties. It is defined as:

$$H = -\sum_{i=1}^{n} p_i \log_2 (p_i),$$

where p_i is the probability that the discrete random quantity considered takes the value x_i. Computing the Shannon entropy for these examples gives $H = 2$ in case (2) and $H = 0.78$ in case (3), showing that the uncertainties are considerably larger in the former case than in the latter.

Using the Shannon entropy, the uncertainties for case (1) are the same as in case (2), but are they comparable? In case 1 we have known frequentist probabilities, whereas in case (2) the probabilities are subjective. In case (1) we know how X will vary when we perform a large number of repeated experiments. This is in contrast to situation (2), where we have the same distribution of X but there is no 'guarantee' that this distribution would be the true one if repeated experiments could be carried out. In case (2) we do not in fact presume that such repeated experiments can be conducted; the outcome of X may be a unique situation in contrast to (1), which per definition presumes the existence of such repeated experiments.

The probabilities in (2) may be based on frequentist probabilities as in (1), but the same numbers could also be assigned when the information/knowledge is much weaker. While probabilities can always be assigned, the origin and amount of information/knowledge supporting the assignments are not reflected by the numbers produced. Hence, we cannot replace uncertainties simply by probabilities. The probability tool is not fully able to represent the uncertainties. Many existing structures for characterising the uncertainties are based on probabilities; see bibliographic notes and Appendix A.

The uncertainty component covers not only the representation Q but also the background knowledge K; i.e. the uncertainties are reflected by the pair (Q,K) and relate to the defined consequences C'. If Q = P this means that a high score for the uncertainties is not only associated with the probabilities of the occurrence of some specific consequences, but is also dependent on judgements about the knowledge supporting the probability assignments. If we judge the background knowledge to be poor, this would affect the total score of the uncertainties in terms of whether they are high or low. Aspects to consider when making judgements about K being strong/poor are (see also Section 4.2):

- The degree to which the assumptions made represent strong simplifications.
- The availability of relevant data.
- The degree of agreement/consensus among experts.
- The degree of understanding of the phenomena involved.
- The existence of accurate models.

Returning to the above example, we may conclude that the uncertainties would be judged to be relatively low in case (1), where a probability model (with no uncertainty parameter) providing accurate predictions of the distribution of outcomes of X if repeating the experiment over and over again is justified. In cases (2) and (3) the uncertainties may be judged to be high, provided that the subjective probabilities are not supported by a 'qualified' prediction model or a probability model.

In case (1) we have assumed that the frequentist probabilities P_f are known. If this is not the case, estimates of P_f have to be produced and the overall judgement of the uncertainties U needs to reflect the level of the parameter uncertainties.

As we have seen from the above, describing/measuring the uncertainties is not straightforward, and many struggle, even within expert societies and groups. A good example is the ISO (2009a, b) standard on risk management, which refers to 'likelihood' in this context but fails to provide a meaningful definition and interpretation of this term.

The analysis has clearly demonstrated the challenges of reflecting the 'magnitude' of the uncertainties. Of course, probability is a key tool, but it cannot fully describe the uncertainties. Interval probabilities represent one possible method, but qualitative approaches also seem adequate for this purpose, to reflect the strength of the background knowledge.

3.3.3 Uncertainty about a phenomenon

Uncertainties about an unknown quantity X and the future C are in some sense easy to define, as there are assumed to be some true values of X and C, although there could be measurement problems. The situation is not so clear when it comes to uncertainties about a phenomenon, such as relevant cause–effect relationships – for example, between smoking and lung cancer, or between extensive use of mobile phones and brain tumours. Another example is the phenomenon of leakages on an oil and gas installation. What are the uncertainties in these cases? Let us assume for the sake of the analysis that a true relationship exists between Z and X, where Z is the output quantity of interest and X is a vector of explanatory (cause) quantities. Hence, a true function g exists – such that $Z = g(X)$ – and we may define uncertainties about g as this true function is of course in general unknown. Uncertainty about the phenomenon is then reduced to a problem wherein some underlying correct value exists, and we can use probability (or other tools) to express uncertainties. Say that we consider two functions g_1 and g_2, for which we may assign probabilities of 0.2 and 0.8 respectively, and we need to take account of the strength of the background knowledge that supports these judgements. We see that we are back to the set-up studied in Sections 3.3.1 and 3.3.2, for a quantity and the future respectively.

Unfortunately the concept of the true function g in this case is not so easily interpreted. This issue has been thoroughly discussed in the literature, with many researchers rejecting the concept that there is in fact a true cause–effect relationship. The argument concerns models linking different quantities. Of course, these models may be more or less good at explaining the phenomena studied, but no model is the true one. The perspective taken in this book is in line with this view.

For smoking, we have strong phenomenon knowledge and we have established a lot of accurate models for predicting fatalities based on smoking. If Z is the number of deaths per (say) 100,000 persons (lung cancer mortality rate) in a population (for example, women of a specific age group), accurate models can be derived linking Z and the intensity X_1 (number of cigarettes per day) and duration of smoking X_2 (years) using standard statistical analysis; see for example Flanders *et al.* (2003) and Yamaguchi *et al.* (2000). We all know that correlation between various quantities does not prove cause-effect relationships, but certainly there is extremely strong evidence showing that smoking increases your lung cancer risk, and the more you smoke, the higher the risk. Here, risk is to be interpreted as the frequentist probability associated with the population you belong to (age group, gender).

It should be emphasised that accurate prediction models can to a varying degree be causal. Many quantities can be predicted accurately even without any deeper understanding of causation. Consider the following statistical model:

$$\mu = 5 - 0.5a \qquad (3.3)$$

where μ is the average of the residual lives of a population of components, and a is the age of the component. Now suppose that a more precise description of the causal relationship is given by

$$\mu = 5 + 0.5a - b$$

where b is a cumulative shock quantity (which is in fact equal to a). The example demonstrates that a prediction model (here [3.3]) can provide accurate predictions (of μ) despite the fact that we have a poor understanding of the cause–effect relationship. A model that we use could be more or less good at explaining the phenomena studied, and its usefulness must always be seen in relation to its purpose. A crude model may be adequate for some applications, but fail to provide insights and decision support in others.

Hence, uncertainty in relation to phenomena is about studying how Z depends on various underlying quantities X, being able to understand and assess uncertainties about Z in view of different models g_i and input on X. For a given model g_i, we are thus back to the problem of assessing a quantity where a true value exists, namely the model error $Z - g_i(X)$; see Section 3.3.1.

Let us return to the example of leakages on an oil and gas installation. Considerable knowledge exists concerning how leakages occur; see Table 3.1. However, there are always some phenomenological uncertainties. Over time the production leads to changes in operating conditions, such as the increased production of water, H_2S and CO_2 content, scaling, bacteria growth, emulsions, etc. – problems that to a large extent need to be solved by the addition of chemicals. What are the effects on corrosion and material brittleness? We need to make judgements about the effects and assess uncertainties. If we use such and such chemicals, what will be the consequences? We are back to the problem of assessing uncertainties about the future, conditional on some specific risk sources or exposure. These assessments are linked to cause–effect relationships as discussed above, and may be reflected to varying degrees in the prediction models of the number of leakages to occur in coming years.

3.3.4 Whose uncertainty assessments (probability distributions) does a risk assessment report: the analysts' or the experts'?

In practical risk assessments there will often be a separation between the experts and the analysts, and *it is a common conception that it is not the analysts' beliefs that matter but solely the experts'* (O'Hagan and Oakley 2004); the analysts are there to facilitate the elicitation of the experts' knowledge and beliefs, synthesise it and report the results as integrated uncertainty assessments, for example expressed through a set of probability distributions. Analysts' judgements beyond these tasks should not be incorporated in the uncertainty assessments (distributions).

This conception that probability assessment results represent only the experts' beliefs and not the analysts' can be challenged, and it is obviously linked to the discussion in Section 3.3.1 concerning the aim of the risk assessment (items (1) and (2)). Using two examples, we demonstrate that the results of the risk assessments are strongly influenced by the analysts, even if the objectives of the assessments could be to faithfully represent and report the expert knowledge. This influence arises from both the analysts' knowledge of probability and the assessment process, *and* their knowledge of the subject matter of the assessment process. However, the two cannot be separated. The analysts' subject matter knowledge or lack thereof influences the choices they make about the assessment process. The problem is simply that the assessments cannot be conducted without the active involvement of the analysts. For example, the analysts have to determine the models to use and how the experts should express their knowledge, and these tasks affect the results to a great extent. The first example we consider relates to the number of events of a certain type occurring in a specified period of time. Several experts are used to provide input to the assessment. The second example is a large quantitative risk assessment – QRA (probabilistic risk assessment – PRA) – of a process plant. This assessment is much more complicated than the first one since it involves extensive use of models comprising a large number of parameters. In both cases the data are scarce.

THE OCCURRENCE RATE EXAMPLE

We are studying the possible occurrences of events A during a time interval [0,t] in the future. The number of events is anticipated to be rather few, and there is not much relevant data that can be used for predicting this number. As concrete examples we may think of the events as failures of a new type of technological system or a system not used frequently, or terrorist attacks. A risk assessment is to be conducted to support decision-making on how to manage the risk and uncertainties related to the possible occurrences of these events. The decision-maker consults a group of risk analysts to perform this assessment. The task is to provide a faithful representation and report of the risk and uncertainties based on the available knowledge. A number of experts in the specific fields studied are to be used to describe this knowledge. The result of the assessment is the generation of a set of probability distributions of unknown quantities.

The issue we will discuss is: do the probability distributions reported reflect the uncertainties (degrees of belief) of the risk analysts, of the experts, or of a combination of these groups? How should the results of the assessment be reported to the decision-maker in order to faithfully describe the process leading to the final probability distributions?

To answer these questions, we need to look more closely into the probability assignment processes. We have to identify precisely what the risk analysts do in these processes and relate them to the type of information and knowledge the experts provide.

The main tasks of the risk analysts are:

(1) To formulate in professional risk assessment language what the purpose of the assessment is, i.e. what type of information should be reported to the decision-maker.
(2) To develop the methods that should be used for assessing the risk and uncertainties.
(3) To consider the need for developing models of the phenomena studied and develop such models if appropriate for the purpose of the assessment.
(4) To identify which experts should be used in the assessments (if not determined by the decision-maker) and how information from these experts should be elicited and, possibly, aggregated.

These tasks are highly integrated; for example, the method and models used are closely linked to the purpose of the assessment. Let us indicate how the risk analysts may conduct these tasks. As we will see, there is no single obvious solution.

First, the quantities of interest need to be specified. Essentially, there are two types of quantities that could be addressed in the assessment:

(1) X = the actual number of events A occurring in the time interval [0,t], or, equivalently, X/t, the average number of events occurring per unit of time during [0,t].

(2) λ = the occurrence rate of A per unit of time, understood as the number (or the expected number) of events A occurring per unit of time in the long run when assuming that the likelihood of events occurring does not change over time. Alternatively, λ may be interpreted as the average number of events occurring during a unit of time when considering an infinite number of similar time units.

The introduction of λ means that a probability model is introduced that reflects the stochastic (aleatory) uncertainties, i.e. the variation in the numbers of events occurring during different periods of time. The common model for the problem studied is a Poisson model, where λ is the intensity of the process.

The probability model represents a judgement on the part of either the analysts or the experts, and it needs to be justified. Does it make sense to consider repetitions of the activity or system studied? Think of the two examples mentioned above: a new type of technological system and terrorist attacks. For the technological system, it is possible to think of a huge population of similar systems, but for the terrorism example the situation is unique. Hence, it would be difficult to justify a probability model in the latter case. Nonetheless, many analysts simply assume the existence of such a model and focus on the statistical inference of the parameters of the model, implicitly introducing their own judgements into the results of the assessment process. This approach avoids explicitly addressing the epistemic uncertainty inherent in the rate model.

The next task is to express the uncertainties about the unknown quantities. Depending on the introduction or not of a probability model, these quantities are either X (or equivalently X/t) or λ. The risk analysts may have a general competence in the field studied, but in this case we may assume that experts exist who have more in-depth knowledge about the system or activity being studied. These experts are consulted to strengthen the basis for the probability assignments to be produced. Different approaches can be used for the elicitation of knowledge from the experts, including the following:

(a) Let the experts report all available evidence about the event considered. Seek one integrated probability distribution for the whole group of experts. The analysts may be part of this group if they are experts in the field considered.

(b) Elicit probability distributions for each expert and use a procedure for integrating this into one expert distribution, including the assessment of necessary information about the dependencies between experts.

(c) Elicit probability distributions for each expert and consider these distributions as input to an overall risk analyst probability distribution, in which the risk analyst somehow combines the experts' distributions (and possibly other relevant information) into a single distribution.

(d) The same as (c), but report both the expert distributions and the risk analyst distribution.

(e) Ask the experts for less information than a probability distribution (for example, high and low values, or a 90% uncertainty interval and the median/expected value). Report only these high and low/interval values, possibly using them for sensitivity analysis in the risk assessment.

These are just examples of the many possible ways that knowledge can be elicited from experts and combined with analysts' judgements. The implications of these different approaches will be discussed later, after the second example is introduced.

THE QRA EXAMPLE

The second example considers the design of a petroleum process plant. To support decision-making about the choice of arrangements and risk-reducing measures, a risk assessment is conducted. A group of risk analysts is consulted to produce the decision basis. Several experts on specific issues, such as explosions and fires, are consulted to provide input to the QRA. The task of the assessment is to provide a faithful representation and reporting of the knowledge (and lack of knowledge) available concerning the system analysed. The result of the assessment is the generation of a set of probability distributions of unknown quantities.

The problems raised are similar to those considered in the previous example, and the four tasks of the risk analysts (1 to 4 in the list above) are the same. In the following we give a brief summary of the main features of the risk assessments, and, as above, there are different approaches for how to solve the problems. For the sake of simplicity we restrict our attention to third party risks, i.e. risk caused by events at the plant affecting people living, working or staying outside the plant.

The quantities of interest first need to be determined. Again the focus could be on the "observables" or parameters of probability models. In the former case the interest would be on quantities such as:

N: the number of fatalities (third parties).

D: the occurrence of an accident leading to a fatality of person z arbitrarily chosen from the surrounding population.

A main aim of the risk assessment is to predict these quantities and to describe uncertainties. In this case the predictions would be straightforward: there would be no fatalities and the event D would not occur. However, there are uncertainties and accidents could occur, possibly leading to deaths.

To describe these uncertainties, event tree models and other physical models are introduced; in addition, probabilities (subjective, knowledge-based) are used to express the uncertainties.

In the latter case the interesting quantities are frequentist probabilities (chances):

> p = individual risk for a specific person in the group having the highest risk, i.e. the frequentist probability that a specific person (arbitrarily chosen) shall be killed due to the activity during a period of one year

and

> p_n = the frequentist probability that an accident will occur leading to at least n fatalities.

The further assessment of these parameters can be conducted in different ways. Two of the most common are:

(i)   Estimates p^* and $(p_n)^*$ of these frequentist probabilities are derived, using the models developed, hard data and expert judgements. No uncertainty assessment beyond these estimates is carried out.
(ii) A full probability of frequency analysis is performed (Kaplan and Garrick 1981), i.e. subjective probabilities are used to express epistemic uncertainties about the parameters p and p_n. Again, the basis comprises the models developed, hard data and expert judgements.

Bayesian analysis may be used for updating the subjective probabilities to formally incorporate new information. If such analysis is conducted, the analysts must carry out the following steps:

•   Assign prior distributions on the parameters of interest.
•   Use Bayes' theorem to establish the posterior distribution of the parameters.

The probability model in (i) and (ii) would express that $p = g(q)$ and $p_n = g_n(q)$, where g and g_n are functions of a set of parameters $q = (q_1, q_2, ...)$, representing, for example, the frequentist probabilities (chances) of some branching events of an event tree or the basic event of a fault tree. Other approaches may also be applied, for example using intervals for some of the parameters q, as was discussed in the first part of Section 3.3.1.

To provide estimates of the parameters q and/or to assign epistemic uncertainties about q, expert judgements are often used, at least for some of the parameters. As in the first example, different approaches can be used for the elicitation of knowledge from the experts, including those mentioned above (a to e).

DISCUSSION

The previous section has outlined ways in which risk assessments can be carried out. The aim of the risk assessment is to faithfully represent and report the knowledge available, but we have seen that there is no single approach that can be used to meet this aim. The role of the experts varies, and the common conception that it is not the beliefs of the analysts that matter but only those of the experts may be challenged. Let us look more closely into this thesis by first considering the QRA example.

The key question we ask is: whose uncertainty does the risk assessment express? Our answer is that it is the uncertainty of both the risk analysts and the experts, but filtered through the lens and experience of the risk analysts' choices and assumptions, as argued in the following.

First, let us look at the case where the risk assessment simply produces best estimates p^* and $(p_n)^*$ for the parameters p and p_n. These estimates are a result of judgements made by both the experts and the risk analysts:

- the derivation of the models g and g_n (risk analyst judgement with expert involvement or at least confirmation);
- the identification and selection of hard data used to estimate selected parameters q_i (primarily risk analyst judgement with expert confirmatory examination);
- the choice of experts and the method used to elicit the expert knowledge and combine different sources for estimating selected parameters q_i (primarily risk analyst judgement).

A common approach used to estimate the parameters q_i is to conduct "analyst judgements based on all sources of information" (Hoffman and Kaplan 1999). This method is appropriate when data are absent or are only partially relevant to the assessment endpoint. The responsibility for summarising the state of knowledge, producing the written rationale and specifying the probabilities rests with the analysts.

For some parameters q_i the estimates could be based on expert judgements alone, but in many cases the number of such parameters is few compared to those estimated by the risk analysts themselves, mainly through the method of "analyst judgements based on all sources of information". Time and resource constraints would not allow for too many expert-assigned estimates. Important exceptions are large and prominent well-funded studies such as the risk assessment for the high-level nuclear waste repository in the US. In such cases, the aim is to base parameter estimates primarily on expert knowledge via formal elicitation procedures.

An important distinction to be made in the types of knowledge used in the expert elicitation discussed here is that of subject matter knowledge versus model and elicitation knowledge. Subject matter experts possess deep knowledge of a fairly narrow field, and generally have far more knowledge than risk analysts possess in the particular realm of enquiry. However, they generally have limited formal training and knowledge about the elicitation process and the overarching risk model. This is the expertise of the risk analysts. It is natural then for probability assessments to be founded on the knowledge of subject matter experts, but to be heavily filtered through – and in some cases influenced by – the expert knowledge of the risk analysts. The subject matter expertise forms an important starting point, but the risk analysts' knowledge heavily influences the end result. Furthermore, the risk analysts' subject matter knowledge can and often does influence the assessment process. In some cases this is direct and explicit as in the "analyst judgements based on all sources of information" discussed above. In other cases the influence is much more subtle. A risk analyst has many decisions to make about how to conduct a given assessment, including which experts to choose, how to structure the assessment protocol, which specific assessment methods to use, and which, if any, starting values to supply for the assessment. Their choices on decisions such as these influence the outcome of the assessment, and these choices are made based on the analysts' knowledge of assessment processes, the subject matter of the assessment, and the particular experts potentially involved.

Now, if the assessment is carried out according to the probability of frequency approach (a Bayesian approach), epistemic uncertainties of the parameters need to be expressed through (subjective) probabilities. Expert judgements are then strongly required, and they can take different forms, as (a) to (e) listed above. What we have said for the estimation case (p^* and (p_n)*) also applies to the probability of frequency (Bayesian) approach. While the expert knowledge forms the foundation of the end result, it is heavily filtered through and influenced by the risk analysts' knowledge. In addition, the time and resource issue can be particularly important in this case since there are so many parameters for which uncertainty distributions are to be determined. If insufficient resources are available for full expert-based elicitation for all parameters, risk analyst knowledge is sometimes substituted for those parameters for which a full elicitation is not conducted. The number of different combinations of experts that could be consulted is also an issue here. If the assessment was to report solely the experts' knowledge, it would be difficult to argue that a particular combination x of experts is the one to use in preference to the different combinations y or z.

An additional issue arises for situations in which a formal Bayesian updating process is to be performed. In order to conduct a formal Bayesian updating process prior distributions have to be assigned, for example using one of the policies (a), (b) or (c) mentioned above. A so-called non-informative prior distribution is often suggested in order to avoid judgements being made that extend beyond the information provided by the data and the experts. A distribution commonly used for this purpose is uniform distribution. However, as discussed in Section 3.3.1, if a uniform distribution for [0,1] is assigned, the analysts/experts have assigned the same probability (1/2) for the quantity to be in the interval [0, ½] as in [½, 1]; this is a judgement made by the analysts/experts, and it cannot be seen as non-informative.

The same type of problems occur if we try to see the risk assessment results as representations of experts' knowledge in the case where the focus of the QRA is on the 'observables', such as the number of fatalities, and no probability model is introduced to reflect stochastic (aleatory) uncertainties. The difference is just that we obtain probability distributions of the observables and not of the parameters of the probability models. The latter assessment is much more complicated than the former since it is based on two layers of probabilities: frequentist probabilities (chances) and subjective probabilities. To a large extent the approach chosen influences the way the uncertainties are represented and reported – this choice is completely determined by the analysts.

Finally, we address the analyst–expert issue when we use approaches for representing uncertainties other than the standard probability-based approaches, such as interval analysis and possibility theory. These approaches are not commonly used, but nonetheless we may discuss the role of the experts and analysts. Clearly, using intervals may eliminate some of the analysts' judgements when few data are available. Hence, the bounds derived for the quantities addressed could be said to be more expert-based than in the pure probability-based approaches. However, the results of the assessments would still be determined by the analysts to a large extent: their choice of which quantities to address, the assessment method and in particular the elicitation procedure. If the risk analysts were to aim at eliciting probability distributions from the experts, the uncertainty representation and report would be completely different from those which would be produced if intervals were sought. The experts may be more or less comfortable with assigning specific probability distributions, but with proper training they would learn how to express their judgements through the probabilities. It is difficult to determine which approach provides the most faithful representation and report of the knowledge of the experts. The point being made is that the risk analysts play the key role in deciding the elicitation procedure and hence the way in which the results of the assessment are expressed.

Next we discuss the occurrence rate example. Again, the key question we ask is: whose uncertainty does the risk assessment express? In the QRA example our answer was that of the risk analysts, but for the occurrence rate example such a conclusion would not in general be justified.

Say that the procedure (a) mentioned above is used for eliciting knowledge from the experts:

(a) Let the experts report all available evidence about the event considered. Seek one integrated probability distribution for the whole group of experts (the analysts may be part of this group if they are experts in the field considered).

Then we obtain integrated probability distributions for the unknown quantities, which to a large extent are determined by the experts. The risk analysts simply facilitate the elicitation process.

However, in this case the analysts also play an important role concerning the way in which the uncertainties are represented and reported. In adopting procedure (a), the aim of the elicitation process is probability distributions. That is a choice that means that the experts need to express their judgements about the unknown quantities. How probable is it that X or λ is below a fixed value? A specific number is required. If, on the other hand, the analysts were in favour of using intervals to represent the uncertainties, the experts' task would be to produce intervals. For example, bounds (imprecision intervals) are established for the probability that λ exceeds a number λ_0, i.e. $P(\lambda > \lambda_0)$, such as

$$0.3 \le P(\lambda > \lambda_0) \le 0.7.$$

The idea is that the assessors (the experts) are not willing or able to express their probabilities more precisely than this. Clearly, such an interval would report the uncertainties differently from that of a specific probability (say 0.5), which expresses the subjective probability of the assessors.

Whether focus of the assessment is on X (alternatively X/t) or λ is also determined by the risk analysts. It makes a lot of difference whether the uncertainty assessments address the number of events A or the occurrence rate of such events. Think of the following two probabilities: $P(X/t > 5)$ and $P(\lambda > 5)$. The former expresses the probability that the average number of events occurring per unit of time in the interval $[0,t]$ exceeds five, whereas the latter expresses the probability that the occurrence rate of events exceeds five, where the occurrence rate is a model parameter interpreted as the average number of events occurring per unit of time in the long run, when assuming that the likelihood of events occurring is not changing over time. Obviously the latter situation is completely different from the former, as there is no restriction on the event-generating process when assigning $P(X/t > 5)$, whereas the assignment of $P(\lambda > 5)$ is conditional on a model assumption.

Nonetheless, one may say that the reported numbers are those of the experts. This is also the case if the elicitation procedure (b) is used (elicit probability distributions for each expert and use a procedure for integrating this into one expert distribution), although the analysts determine the way the different distributions are fused. The fusion can be performed in many ways and the procedure used strongly affects the output distribution.

Of course, if the elicitation procedure is of the form (c), elicit probability distributions for each expert and consider these distributions as input to an overall risk analyst probability distribution, the resulting distribution is that of the risk analysts. The approach is similar to "analyst judgements based on all sources of information" as referred to in the previous section. The responsibility for summarising the state of knowledge, producing the written rationale and specifying the final probabilities rests with the analysts.

If procedure (d) is used, the underlying expert distributions are also reported as results of the assessments. In this way a broader uncertainty characterisation is obtained, but the information is more difficult to interpret.

These two examples demonstrate that the results of risk assessments represent both the experts' and the analysts' knowledge. The produced probability assessments are founded on the knowledge of subject matter experts, but they are heavily filtered through, and in some cases heavily influenced by, the expert knowledge of the risk analysts. The risk analysts' role is critical for the way the uncertainties are represented and reported. One key aspect is the type of unknown quantities focused on in the assessments (either observables or model parameters), while another and even more important aspect is the type of assignments to be made: should the experts express subjective probabilities reflecting their judgements (degrees of belief) concerning the events of interests, or limit themselves to intervals for the probabilities? In the latter case, the assessor is not willing or able to assign specific probabilities for the events considered.

In complex QRAs the risk analysts often play the leading role in the assessments in the sense that the resulting probability distributions are to a large extent determined by the analysts. Expert judgements may provide useful input, but for such risk assessments this input is often marginal compared to the many judgements that the risk analysts have to make to conduct the risk assessment.

Nevertheless, for less complex risk assessments having few parameters (unknown quantities), such as the occurrence rate example, the thesis referred to above can be justified under certain conditions, as we have seen in the preceding analysis. However, also in this case, the risk analysts influence the representation and report of the uncertainties to a large extent, as discussed at the beginning of this conclusion section. It is therefore essential in the reporting of the results of the risk assessments that the premises and the approach taken for the assessment are thoroughly discussed. Referring to alternative approaches that could have been used for the assessments and the expert elicitation in particular would provide useful information for the decision-makers when they interpret the results of the assessments. As a rule, sensitivity analyses showing how the representations and reports are affected by the different approaches should be conducted.

3.4 Surprises and black swans

An event is commonly considered a surprise when it occurs unexpectedly and also runs counter to accepted knowledge (Gross 2010). However, there are many other definitions; for example, a surprising event may be regarded as one whose occurrence was not anticipated, or which has been allocated such a low probability that the possibility of its occurrence was effectively discounted (Kay 1984, p. 69). The literature includes many taxonomies for classifying surprises. Examples include the dichotomies between known (imaginable) surprises and unknown surprises, between unanticipated surprises and anticipated surprises, and between unintended, imaginable and anticipated surprises (Gross 2010, pp. 37–41). An imaginable surprise occurs when the event type is known but its occurrence was considered highly unlikely. If we know that something is going to happen, but not when and in what form, it is referred to as an anticipated surprise. As noted by Gross (2010, p. 40), a surprise cannot be registered in any meaningful way without an 'expectation' in some sense, to create a deviation.

Black swans as defined by Taleb (2007) and others (see Section 1.7) are a type of surprise. In line with Aven (2013a), a black swan is seen as a surprising extreme event relative to the present knowledge/beliefs. Hence the concept must always be viewed in relation to whose knowledge/beliefs we are talking about, and at what time. Building on this definition, Aven and Krohn (2014) distinguish between three types of such events:

(a) Events that were completely unknown to the scientific environment (unknown unknowns).
(b) Events not on the list of known events from the perspective of those who carried out a risk analysis (or another stakeholder), but known to others (unknown knowns – unknown events to some, known to others).
(c) Events on the list of known events in the risk analysis but judged to have negligible probability of occurrence, and thus not believed to occur.

The term 'black swan' is used to express any of these types of events, tacitly assuming that they carry an extreme impact. See Figure 3.7, which links terms such as 'black swans', 'surprising events' and 'unforeseen events'.

The first category of black swan type events (a) is the extreme – the event is unthinkable and/or unknown to the scientific community, for example a new type of virus. In activities about which there is considerable knowledge, such unknown unknowns are likely to be rarer than in cases of severe or deep uncertainties.

The second type of black swan (b) is events that are not captured by the relevant risk assessments, either because we do not know them, or we have not made a sufficiently thorough consideration. If the event then occurs, it was not foreseen. If a more thorough risk analysis had been conducted, some of these events could have been identified. The third category of black swans (c) is events that occur despite the fact that the probability of occurrence is judged to be negligible.

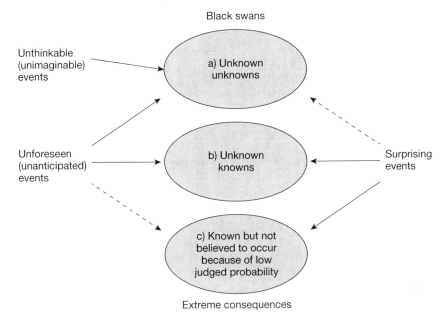

Figure 3.7 Schematic illustration of the concept of black swans: unknown unknowns, unforeseen events, surprising events and unthinkable events, based on the ideas presented by Aven and Krohn (2014) and first presented in Aven (2013g)

An (unanticipated) surprising event (with severe impacts) is thus a black swan according to this logic. Events from categories (b) and (c) will obviously come as a surprise, but this is not so obvious when we talk about category (a), unknown unknowns – therefore the dotted arrow in Figure 3.7. Considering an activity with deep uncertainties about the type of events that will occur and the impact they will generate, we may be completely free of "expectations" for what is coming. Hence, it may be questioned whether an unknown unknown does in fact come as a surprise in such a situation.

Similarly, we may problematise what is an unforeseen event. If an event occurs which was judged to have negligible probability, was it then foreseen? Yes, in the sense that the fact that it could happen was anticipated, but also no, in the sense that it was not considered likely.

Think of a container of fluid. Normally it is filled with water and people drink from it daily. One day Susan drinks fluid from the container and it turns out to be toxic. We refer to this as a black swan, a surprise in relation to her knowledge/beliefs (assuming that it carries a serious impact). To be labelled a black swan, the event need not be a new phenomenon or an unknown unknown. In retrospect, we can easily explain the incident.

A risk analysis could have identified such an event, but nevertheless it may be surprising for some people (particularly Susan) in relation to their beliefs/

knowledge. These are the types of events we are concerned about. Let us modify the example a little. Suppose a risk analysis has identified various types of toxic fluids that could fill the container in special situations, but it excludes a dangerous form because of a set of physical arguments. Then, however, this scenario occurs. The event was possible, despite the fact that it was considered impossible (extremely unlikely) by the analysts. The real-life conditions were not the same as those that were the basis for the risk analysis, and the event came as a surprise even for the risk analysts. In retrospect, however, it was easily explained.

In relation to classifications (a) to (c) above, the event is classed as belonging to category (c). For the first case (Susan), the basis is the beliefs that this person has, and it can then be placed under (b) or possibly (c).

A concrete example of a black swan of this kind is discussed by Sande (2013). Until approximately 1960 it was accepted knowledge that natural gas cannot explode in open air. This knowledge was underpinned by experimental tests, but the gas volumes that were used in these experiments were too small. Later experiments with larger volumes of different types of gases have shown that all hydrocarbon gases can explode in the open. However, different gases have different 'critical diameters', which is the minimum diameter that must be present to allow a continuous detonation.

Strictly speaking, it would make sense to say that an unthinkable event is an unknown unknown. However, we may also argue differently. Viewed from a risk assessment point of view, an event that belongs to category (b) may also be judged as unthinkable provided a thorough analysis has been performed to uncover all relevant events. The question is: unthinkable for whom?

In a risk setting, the idea of unknown unknowns intuitively captures the fact that the actual events occurring are not covered by the events identified in the risk description/risk assessment. Of course, our focus here is on events with extreme consequences. Consider the risk perspective (C, U), reformulated by specifically showing some events A included in C: (A,C,U) (for example, A may represent a terrorist attack or a gas leakage in a process plant). When speaking about the risk (A,C,U), there are no unknown unknowns, as A and C simply express the actual events and consequences of the activity. However, in relation to a risk description (A',C',Q,K), we may have unknown unknowns (here A' and C' are the events and consequences respectively, specified in the risk assessment, while Q is the measure of uncertainty used and K is the background knowledge; see Section 2.4). The point is simply that the A' events do not capture the A; we may experience some surprises relative to A'. For example, if we study the life of a young person, he or she may die of a disease not known today; the A' events do not cover the true A. Hence the unknown unknowns are included in the risk concept, but they are not captured by the risk description. In practice, while the A events may not be known to a specific risk analyst team, they may be known by others. Then it is better to refer to them as unknown knowns, as indicated for category (b) above.

It is common to employ the term 'unexpected' when characterising surprises and black swan events. However, this term is problematic. Consider the following example. An event has three possible outcomes, 0, 50 and 100, with associated probabilities 0.25, 0.50 and 0.25 respectively. Hence the expected value in a statistical sense (the centre of gravity of the probability distribution) is equal to 50. The values 0 and 100 can thus be seen as unexpected; however, the probability of one of these outcomes occurring is 50%, which cannot be viewed as surprising. Clearly for the term 'unexpected' to make sense, we need to interpret it in relation to the probability distribution. What is 'unexpected' needs to be understood more as an outcome not belonging to a sufficiently broad uncertainty interval [a,b], such that the probability of the quantity of interest not being covered by this interval is small, (say) less than 5%.

What is surprising must always be understood in relation to who considers it to be a surprise, and when. Figures 3.8 to 3.11 illustrate this. We consider an activity, for example the operation of an offshore installation at a given

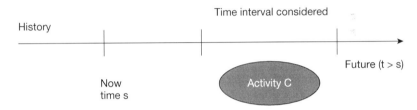

Figure 3.8 Illustration of risk in relation to the time dimension. C: consequence of activity (based on Veland and Aven 2014)

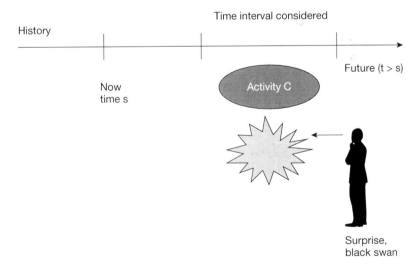

Figure 3.9 Illustration of the relationship between risk, black swan and the time dimension (based on Veland and Aven 2014)

future time period, for instance next year. We let C denote the consequences of the activity in relation to the values with which we are concerned (life, health, environmental, assets). What C will be is unknown to us at time s; there are risks present (cf. the understanding of the concept of risk in Section 2.4). We assume now that a risk assessment of the activity has been conducted at time s. Time goes on, and C is realised, usually without a major accident occurring. However, let us imagine that such an accident actually occurs, as shown in Figure 3.9. It is a result of the occurrence of a combination of events and conditions, and comes as a surprise to those involved in the management of the activity. The accident is a black swan for them. Now let us take a macro perspective – looking at a large number of such activities, for example the whole oil and gas industry. Risk is now linked to the occurrence of any major accident in the industry; where and how the event occurs is not the issue. Again a risk assessment is conducted. It is concluded that there is a relatively high probability that such an accident could occur. Consequently, one cannot say that it is a black swan if such an event actually occurs. See Figures 3.10 and 3.11. From a macro perspective, a realistic analysis would state that we must expect that a major accident will occur somewhere in the next ten years. However, there is no law that says that it will actually happen. We are not subject to fate or destiny. Each unit (organisation, company, installation) works hard to prevent such an accident actually occurring. It is believed that with systematic safety work this goal can be achieved. Accordingly, any such serious accident normally comes as a surprise, a black swan for those involved in the operation and management of the activity.

Paté-Cornell (2012) discusses the concept of black swans and relates it to the 'perfect storm' metaphor. This storm resulted from the combination of a storm that started over the United States, a cold front coming from the north, and the tail of a tropical storm originating in the south. All three meteorological features were known before and occur regularly, but the combination is very rare. The crew of a fishing boat decides to take the risk and face the storm, but they have not foreseen its strength. The storm strikes the boat, which capsizes and sinks; nobody survives (Paté-Cornell 2012).

This extreme storm is now used as a metaphor for a rare event that may occur, where we understand the relevant phenomena. The experts can calculate the probabilities of such events and the associated risks with a high degree of precision. They can make accurate predictions of what will happen, stating that in one in ten such situations the waves will be like this, and in one in a hundred such cases the waves will become so big, etc. When we build oil and gas installations offshore we take into account such events. We set requirements for the installation's strength to enable it to withstand extreme waves, but there is always a limit. We must accept that there may be a wave that is so large that the installation will not tolerate it, but such an event should have a very small probability.

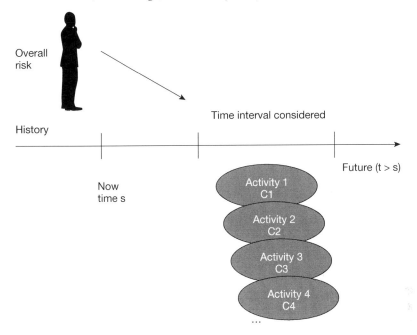

Figure 3.10 Illustration of risk in relation to the time dimension when the perspective is macro, for example the whole oil and gas industry (based on Veland and Aven 2014)

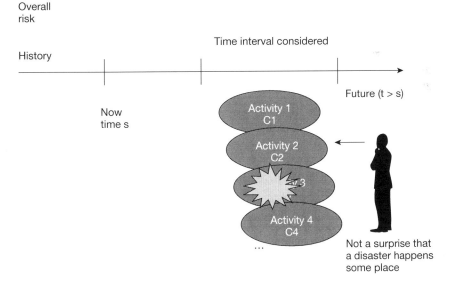

Figure 3.11 Illustration of relationship between risk, black swans and the time dimension when the perspective is macro, for example the whole oil and gas industry (based on Veland and Aven 2014)

The situation has similarities to other areas, such as health and traffic. In many cases we know quite precisely what proportion of the population will contract certain diseases next year, and how many people will die in traffic. Actions can be taken to reduce risks, and we can measure changes over time. When one looks at the number of road traffic fatalities from the 1970s to the present day, the figures show a steady decrease, despite the fact that traffic has increased. The risk management works.

The metaphor of the 'perfect storm' is thus about events where science in its traditional form prevails, where we have precise probabilities and relevant statistics, and where we can make accurate predictions about the future. Black swan events of type c seem to be covered by rare perfect storm events. However, there is an important difference. In relation to perfect storms, the variation in the phenomena is known and we face risk problems where the uncertainties are small; the knowledge base is strong and accurate predictions can be made. As the knowledge base is so strong, black swans can for all practical reasons be ignored. The probabilities are frequentist probabilities, characterising the variation in the phenomena, and they are known to a degree that is viewed as certainty.

For black swans of type (c), we are in a situation where we cannot make this type of accurate prediction. The variation in the phenomena cannot be described with this type of precision. We need to rely on (subjective) judgements, where probability refers to the knowledge-based (judgemental, subjective) assignment of uncertainties and degrees of belief. When stating that an event is judged to have negligible probability and is not expected to occur, it is with reference to such a perspective. Clearly, in such cases we may experience surprises compared to the judgements made.

Think again about the oil and gas installation. The management of the installation may ignore the possibility of a specific event occurring, arguing in this way. It is not a perfect storm type of event as it cannot be predicted with accuracy. Taking the macro perspective for the industry as discussed above, we are closer to the perfect storm situation. If we consider the industry as a whole, it does not make sense to talk about black swans because the probability of occurrence is rather high. However, let us build a thought-construction; we assume that the frequentist probability of the occurrence of such an event is rather low. It is a rare event. Would it then be a black swan? No is the immediate answer, as the variation is still known, the phenomena studied are well understood. Further reflections may however challenge this view.

If we have a situation with perfect information about the variation of the phenomena – we know the frequentist probability distribution (we are in the perfect storm situation), one can argue that the occurrence of a low frequentist probability event should not come as a surprise. It is rare, but it is known with certainty that the event will occur sooner or later. Hence it is not a black swan (type c). However, one can also argue differently. Given the knowledge about the variation in the phenomena, it is considered so unlikely that the event will occur the next year, say, that it is not believed to occur. Hence it can be viewed as a black swan of type (c) if it in fact does occur. Again we see that whether the event is a black swan or not is in the eyes of the beholder.

In practice we cannot fully understand the variation. If we go back some years we would not have thought about terrorism events as a contributing factor to the variation, and hence a black swan may have occurred even though the phenomenon was considered well understood. However, this type of black swan is not of type (c), but of type (a) or (b). The discussion here relates to the distinction between common-cause variation and special-cause variation in the quality discourse, as discussed in Section 3.1. The common-cause variation captures 'normal' system variation, whereas the special causes are linked to the unusual variation and the surprises, the black swans.

We may also talk about 'near-black swans', meaning surprises relative to one's knowledge/beliefs but where the event did not result in extreme consequences; the barriers worked and avoided the extreme outcomes. A black swan can occur as a result of a set of events and conditions, and a subset of these may generate a near-black swan.

LINDLEY'S EXAMPLE

Let us return to the case presented by Lindley (2008), mentioned in Section 1.7. In this example we consider a sequence of independent trials with a constant unknown chance of success. Lindley shows that a black swan (failure of trial) is almost certain to arise if you observe a lot of swans, although the probability that the next swan observed will be white (success of trial) is nearly one. To obtain his probabilities, he assumes a prior probability distribution over this chance, namely a uniform distribution over the interval [0,1]. This means that Lindley has tacitly assumed that there is a zero probability that all swans are white – there is a fraction of swans out there that are black (non-white). From this point on, his analysis cannot change this assumption. Of course, then the probability calculus will show that, when considering a sufficient number of swans, some black ones will be revealed; see Appendix D. Through the assumptions he has made, the analyst has removed the main uncertainty aspect of the analysis. In real life we cannot exclude the possibility that all swans are white. The uncertainty about all swans being white is a key issue here, and Lindley has concealed it in his assumptions. This is the problem raised by many authors: the probability-based approach to treating risk and uncertainties is based on assumptions that could be concealing critical problems, and therefore provide a misleading description of the possible occurrence of future events. Let us reconsider Lindley's example to allow for a positive probability that all swans are white.

Let us assume that there are only two possibilities: the fraction p of white swans is either 100% or 99%. Hence p is either 1 or 0.99. Suppose the analyst assigns prior probabilities to these values as 0.2 and 0.8, respectively.

Now suppose that the analyst has observed n swans and they are all white; what then is his posterior probability for the next m swans to be all white (n and m being large numbers)? Using Bayes' formula in the standard way, we find that this probability is close to one, i.e. the probability of a black swan occurring is very small, in contrast to what Lindley computed in his analysis. See Appendix D for the details.

This example shows the importance of the assumptions made for the probabilistic analysis. Depending on these assumptions, we arrive at completely different conclusions about the probability of a black swan occurring.

Lindley's example also fails to reflect the essence of the black swan issue in another way. In real life the definition of a probability model and chance cannot always be justified, as discussed in Section 2.2. Lindley's set-up is the common framework used in both traditional statistics and Bayesian analysis. Statisticians and others often simply presume the existence of this framework, and the elements of surprise that Taleb and others are concerned about fall outside the scope of the analyses. This is the key problem of the probability-based approach to risk analysis, and a possible interpretation of Taleb's work is the critique of the lack of will and interest among statisticians and others to see beyond this framework when analysing risk.

3.5 Conclusions

The key concepts of this chapter are:

(1) Aspects of the world, quantities of the world (for example, the number of leakages in a process plant).
(2) Models of this world, models of these quantities of the world (for example, an event tree and a probability model representing variation).
(3) Knowledge and uncertainties related to this world, quantities of the world, model parameters and model errors.

The risk analysis field discusses inter alia which aspects and quantities to study, how to model the world, and how to represent the knowledge and uncertainties. A main topic of the chapter is the potential for surprises relative to the knowledge (beliefs). There is a need for extending the traditional distinction between aleatory (stochastic) uncertainties and epistemic uncertainties by adding the surprise (black swan) dimension. Aleatory uncertainties express the variation in a quantity (item 1 above) and are modelled by probability models. The epistemic uncertainties reflect that the "true" values of the studied quantities are unknown, and are expressed using knowledge-based (subjective) probabilities (for example). The black swan risk adds an actor dimension (with time also reflected) to this structure. Whose models and epistemic uncertainties are we referring to? The risk analyst in a particular case may have performed their epistemic assessment on the basis of a belief that the system studied is a standard one, but in real life it may be a special one with completely different properties – a fact well known by other people in the organisation. The modelling may reflect the real world to varying degrees, and the judgements and beliefs of the epistemic analysis could be based on assumptions that are wrong. Figure 3.12 illustrates this issue, and points to the need for incorporating considerations concerning the limitations of the risk assessment of Actor 1, to the left in Figure 3.12, in the broad risk evaluation and the decision-maker's review, as discussed in relation to Figure 3.6.

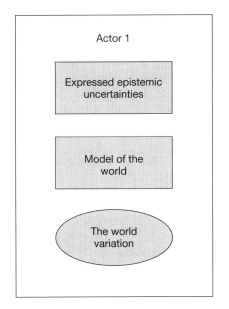
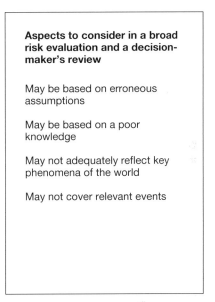

Figure 3.12 Aspects to consider in a broad risk evaluation and decision-maker's review (see Figure 3.6) (based on Veland and Aven 2014)

Bibliographic Notes

Section 3.1 of this chapter is to a large extent based on Aven (2014a). It also draws on some of the philosophical reflections made in Bergman (2009), and to some extent also in Aven and Bergman (2012) and Aven and Krohn (2014). When it comes to quality management, key references are Deming (2000) and Shewhart (1931, 1939). Other main sources for the present work are Bergman and Klefsjö (2003), Cutler (1997), Chakhunashvili and Bergman (2007), Oakland (2003) and Woodall (2000).

It is possible to extend the classification defined by common-cause variation and special-cause variation in many ways. One approach is discussed by Chakhunashvili and Bergman (2007), using the concept of weak statistical control inspired by Shewhart (1931), who makes a distinction between two types of special causes of variation: Type 1, which are sources of variation that do not belong to a system of chance causes and can adversely affect the outcome of a process, and Type 2, which are sources of variation that themselves are composed of chance causes and hence do not necessarily jeopardise the statistical predictability of the process. Chakhunashvili and Bergman (2007) integrate the Type 2 category and the common-cause variation, and obtain a definition of a process in weak statistical control: the process is stationary, or it is influenced only by assignable causes, which are themselves generated by processes in statistical control (or in weak statistical control). However, here our focus is on Type 1 special causes, which are not meaningfully understood to be a process in control.

The main sources of Section 3.2 are Aven (2013a) and Hansson and Aven (2014), as well as basic literature on knowledge and science (e.g. Lemos 2007, Hansson 2013). The DIKW hierarchy as here presented can be traced back to the 1980s, for example Cleveland (1982), Zeleny (1987) and Ackoff (1989), and see also Hey (2012), but the first known registration of the DIKW ideas goes back to T.S. Eliot (1934) and his poem 'The Rock', where he writes:

> Where is the life we have lost in living?
> Where is the wisdom we have lost in knowledge?
> Where is the knowledge we have lost in the information?

The summary of the main features of data, information, knowledge and wisdom in Section 3.2 is based on Frické (2009) (who based his review mainly on traditional sources: Adler [1986], Ackoff [1989] and Zeleny [1987]). Some authors have also referred to a category of "understanding" before wisdom, but for the purpose of the present analysis this has not been considered fruitful. We have partly placed our understanding in the knowledge category as shown in Figures 3.3 and 3.4.

Section 3.3 on uncertainties integrates material from Aven (2010a, b, 2011a, 2013d, e) and Aven and Guikema (2011), in addition to fundamental works such as Paté-Cornell (1996) and Winkler (1996). The paper of Ferson and Ginzburg (1996) is a point of departure for the discussion in Section 3.3.1. For the various schemes for elicitation of information from experts mentioned in Section 3.3.1, further details can be found in Clemen and Winkler (1999), Hoffman and Kaplan (1999) and Cooke (1991). The discussion about the accuracy of prediction models and its links to causation is inspired by Cox (2011) and the discussion in Aven (2011e). For some alternative structures for categorising uncertainties in a risk context, see Walker *et al.* (2003) and Spiegelhalter and Riesch (2014).

The final Section 3.4 is to a large extent built on Aven (2013b, 2014a, c) and Aven and Krohn (2014).

4 Risk assessment

In Chapters 2 and 3 we have established a theoretical basis for understanding and describing risk. Now we will look at some practical implications for risk assessment. We will study how the new risk perspectives change the way that risk is assessed in real-life situations, giving special attention to how we can incorporate the knowledge and surprise dimensions. Several cases will be studied.

Establishing suitable ways of representing and treating the knowledge and surprise dimensions in risk assessment is a huge research topic. The final answer on the issue will not be provided here. Rather, the aim is to point to some key challenges and provide some preliminary ideas and reflections on possible routes for research and development within these areas. Nevertheless, in a few specific areas some quite detailed work and suggestions for how to assess risk are provided.

Traditionally risk assessment methods are divided into two main categories: qualitative and quantitative methods. In this book we distinguish between qualitative methods and semi-quantitative methods only, as any quantitative method should be supplemented with qualitative judgements linked to the knowledge on which the quantitative analysis is based.

Many types of traditional risk assessment methods address the issue of what can happen, for example HAZOP (HAZard and OPerability Studies), HazId (Hazard Identification), fault tree and event tree analysis (Aven 2008, Zio 2007, Meyer and Reniers 2013). Hazardous events and scenarios are identified using these methods, but because risk is an issue, uncertainties and likelihood related to these events and scenarios are also addressed in a qualitative or quantitative way. Based on considerations of probability, some events and scenarios can be judged to pose a negligible risk and do not need to be followed up any further. With an increased focus on knowledge, surprises and black swan events and scenarios, we need to reconsider the use of these methods. We have to challenge the premises on which the analyses are based, the assumptions on which the probability judgements rely.

4.1 Qualitative analysis. Identifying hazards/threats, causes and consequences

As mentioned above, there are a number of methods that can be used for identifying hazards/threats, causes and consequences. They are well-documented and described in the literature and there is no need to repeat them here. A review and discussion of some selected approaches will, however, be provided, as these are considered to be of special interest in the present context which focuses on knowledge and surprises. We start by looking into the anticipatory failure determination (AFD) method.

4.1.1 Anticipatory failure determination (AFD)

Kaplan *et al.* (1999) present an approach for hazard/threat identification and analysis that deserves some attention: the so-called anticipatory failure determination (AFD) method. This is an application of I-TRIZ, a form of the Russian-developed Theory of Inventive Problem Solving, and it is particularly suitable for scenarios involving human error, sabotage, terrorism and the like. The relevance of the TRIZ methodology to risk analysis is rooted in the fact that revealing and identifying failure scenarios is fundamentally a creative act, but it must be carried out systematically, exhaustively, and with diligence (Kaplan *et al.* 1999). Traditional failure analysis addresses the questions: "How did this failure happen?" or "How can this failure happen?". AFD and TRIZ go one step further and pose the question: "If I wanted to create this particular failure, how could I do it?" The power of the technique comes from the process of deliberately 'inventing' failure events and scenarios (Masys 2012).

AFD has two types of applications: AFD-1 and AFD-2. AFD-1 applies to finding causes, i.e. explanations, of a failure that has already occurred. It is called a failure analysis. AFD-2, on the other hand, is a failure prediction analysis, searching to identify possible failures that have not yet occurred. Here we focus on AFD-2. To illustrate the method we will return to the talk example described in Section 1.2.

AFD-2 comprises ten steps. **Step 1** is a formulation of the *original problem*, which in general can be stated as finding *all* (i.e. all of the important) possible failures and failure events in the system at hand.

Step 2 covers a description of the *success scenario*, S0, of the system in terms of the phases of the process and the results achieved at the end of each phase. In our example, the speaker's goal is to give a brilliant talk, in the sense that the audience listens with great interest to what he says and enjoys the way he communicates his message. He also requires a good feeling throughout the whole talk, a sensation characterised by high confidence and "having the audience in the palm of his hand". In addition, he may state subgoals or subresults linked to three phases, the introduction, main content and conclusions, as stated in Table 4.1.

Table 4.1 Desired results (subgoals) in the talk example

Phase	Desired results (in addition to general ones relevant for all phases)
Introduction	The talk is placed in a proper context, the audience become curious about what is coming
Main content	The talk has a substance that is interesting and important for the audience
Conclusions	The audience find the conclusions to arise naturally from the two first phases, and understand the message provided by the conclusions

Step 3 formulates the *inverted problem*, which is to create or produce all the possible ways that failures can occur during these phases. The actual formulation is carried out in the coming steps.

In **Step 4**, all the obvious possible failures of the system that can readily be thought of are written down. To stimulate this activity, the possible *initiating events* (IEs), *harmful end states* (HESs), and *mid-states* (MSs) are focused on separately. These events/states are then combined to generate complete risk scenarios (Sis), which may be structured by means of scenario trees – typically using event trees and fault trees, for example.

In our example, some obvious failures are:

the speaker shows no enthusiasm (IE1), the arguments used in the talk are not valid (IE2), the slides are confusing (IE3), and the message is not clear (IE4).

Examples of obvious harmful end states would be:

the speaker is dissatisfied with his performance (HES1), the speaker is embarrassed (HES2), the speaker's reputation is damaged (HES3), the audience is bored (HES4) and the audience is disappointed (HES5).

From these events we can specify some obvious scenarios, such as:

IE1 → HES4

IE2 → HES3

IE4 → audience confused → HES5

Here we have introduced as mid-state (MS), 'audience confused'. This fourth step can be carried out using standard methods such as HAZOP, HazId, or fault tree and event tree analysis, but already at this step of the analysis the basic idea of AFD can be used, thinking about ways to create failures.

Next we need to move beyond the obvious, as **Step 5** asks us to conduct a survey of the resources available in or around our system that might be useful in creating failure scenarios.

In Kaplan *et al.* (1999), categories of resources that might be present are listed for typical engineering applications (substance, field/energy, space, time, functional, change, systemic, organisational, control devices, protection systems). Specific checklists of resources need to be generated for different types of settings. The idea of this resource review is that it stimulates our thinking about failure modes and failure scenarios that might occur which we had not previously thought of.

For the talk example, types of resources could be technical tools for the presentation, the physical and mental capacities of the speaker (for example, voice, and ability to show and control feelings), and the competence of the audience. We quickly see that such a listing of resources can indicate potential failures, for example related to failure or poor use of equipment, voice, etc.

Step 6 extends the resource checklists to stimulate possible failures, covering aspects linked to our knowledge base for the system, such as:

(a) Typical weak and dangerous zones in a system.
(b) Typical functional failures.
(c) Typical harmful impacts on systems (humans included).
(d) Typical life cycle stages of technological systems.
(e) Typical dangerous periods in system functioning and evolution.
(f) Typical sources of high danger.
(g) Typical disturbances in flows of substance, energy and information.

In our talk example, an especially dangerous period is the opening of the speech (linked to (e)). Clearly, if the speaker's first words indicate that he has a poor background in the subject to be addressed, the audience is likely to lose interest in the talk immediately. We have an example linked to (f) if the audience is highly competent in the field addressed, perhaps more competent than the speaker. A possible failure event then would be that the audience discovers a failure in the speaker's argumentation that he cannot easily grasp given his knowledge, even when it is presented to him in a very clear way by a member of the audience.

For all analysis steps, any scenarios that arise should be adequately numbered and included in suitable scenario trees.

In **Step 7**, the method shifts to an 'incoming tree' point of view with respect to the important HESs and MSs that have been identified. In previous steps we have gone back and forth in our thinking between the questions: "What physical effects or principles can create the desired failure?", "What resources do I need to implement this principle?" and "What resources do I have?". Now we go one step forward and consider additional ways by which these events can be created. We apply the Algorithm for Inventing Problem Solving (ARIZ).

A cleaning of all events and scenarios is also required, to make them understandable. The result is a set of Sis for our problem.

An ARIZ for AFD consists of the following steps:

(1) The general way to produce the desired effect is:
The resulting secondary problem is:
(2) The ideal conditions for realising this harmful effect are:
(3) The known way to provide the ideal conditions is:
(4) The way to change the system is:
(A) – Limitations to providing the ideal conditions are:
(B) – Contradiction – There is a way to produce the harmful effect but it cannot be realised for the following reason:
(C) – According to the Separation Principles, this contradiction may be resolved in the following way:

In our talk example, consider the end state of the speaker being embarrassed (HES2). The speaker is usually well prepared and controlled in his way of speaking, so a scenario where he becomes embarrassed is not easily foreseen. Thus, the secondary problem here is to cause him to be embarrassed. The ideal solution would be that he says something that is wrong or inappropriate and that this is focused on by people in the audience. A number of scenarios can lead to this situation, for example:

- The speaker becomes too confident and starts to take risks, saying things that he did not plan to say.
- He has camouflaged difficult issues, and members of the audience reveal this.
- There are people in the audience that are hostile to the speaker and would like to embarrass him.
- The speaker says negative things about somebody and it turns out that he/she is present (or some close friends or relatives are).

A special innovation guide has been developed for typical engineering applications, helping the user find a way to produce or apply the most popular technological (physical or chemical) effects (Kaplan *et al.* 1999). In general, the guide assists us in answering two kinds of questions:

- How to provide a required result ("I know what should be done, but I don't know how to do it").
- How to apply an available effect (energy or process) in some way other than that which it presently performs.

Then we come to item (4). Firstly, what are the barriers and obstacles for obtaining the ideal conditions, (A)? The speaker has strong knowledge in his field and in general he has a cautious attitude. When it comes to contradiction,

B, the speaker is normally in good control of his feelings, but he still needs to be challenged on some issues that affect him strongly. Applying the I-TRIZ Separation Principles, this contradiction can be resolved by the principle of *Separation in Time*. For example, a member of the audience may refer to what the speaker wrote in a paper some years ago and point to some logical inconsistencies in relation to what he is now saying. The speaker is not prepared for such a comment and is not able to provide a plausible explanation for these inconsistencies.

Step 8 is about suggesting ways to 'worsen' the harmful effects – by intensifying them or keeping them hidden until they become appropriately severe. In our example, the intensification can be obtained by lampooning other researchers' work and the masking by not addressing sensitive issues. This worsening can also be achieved by combining various harmful effects. In our talk case we can think of the audience being bored (HES4) and also disappointed (HES5), for example as a result of the speaker talking with no enthusiasm and lacking a clear message, even though he is a well-recognised researcher so their expectations were high. Another intensifying mechanism is the development of a 'chain' of harmful effects. The speaker may firstly disappoint the audience by skipping an interesting issue, and then they become bored by his lack of ability to create enthusiasm about other issues.

In **Step 9** the revealed harmful effects are analysed. The various scenarios can be presented using different types of trees as discussed above.

The last stage, **Step 10,** concerns the prevention or elimination of identified failure scenarios. In the AFD approach, different concepts for eliminating the scenarios are suggested in line with the I-TRIZ, including:

(1) Eliminate the 'causes', i.e. the conditions that 'cause' the undesired action.

The undesired action may in our case be to skip a specific, sensitive issue, and the elimination of the cause could be clarifying work in advance of the talk in order to have a solid basis for a discussion on it.

(2) Introduce a process that eliminates or reverses the effect of the undesired action.

An example here would be to counteract the situation by means of another action. In our case the speaker could take the action of openly presenting the problems linked to the sensitive issue and stating that the issue needs further analysis.

4.1.2 *Other approaches and methods*

The qualitative analyses can be supported by different types of analysis frameworks, for example Actor Network Theory (ANT) (Latour 2005,

Masys 2012). ANT seeks to understand the dynamics of the system by following the actors – it asks how the world looks through the eyes of the actor doing the work. As highlighted by Dekker and Nyce (2004, p. 1630) and Masys (2012), through this approach issues emerge pertaining to the roles that tools and other artefacts (actors) play in the actor-network in the accomplishments of their tasks.

The area of scenario analysis can add valuable input to the risk assessments. Here, scenarios are developed describing potential future conditions and events, using various techniques (Chermack 2011). There is no search for completeness and characterisations of the uncertainties and risks, as in traditional risk analysis, but in the case of large uncertainties and a lack of accurate prediction models, the generation of such scenarios may provide useful insights about what could happen and possible black swans. The deductive (anticipatory, backwards) scenarios are of particular importance in this respect, where we start from a future imagined event/state of the total system and question what is needed for this to occur. System thinking, which is characterised by seeing wholes and interconnections, is critical if we are to identify black swans, as for example highlighted by many scholars of accident analysis, organisational theory and the quality discourse (Turner and Pidgeon 1997, Deming 2000). For example, using an event tree to reveal scenarios has strong limitations, as the analysis is based on linear inductive thinking (Leveson 2011, Hollnagel *et al.* 2006).

Another method that may be useful in revealing potential surprises and black swans is *red teaming*, which serves as a devil's advocate, offering alternative interpretations and challenging established thinking (Masys 2012). For example, "businesses use red teams to simulate the competition; government organisations use red teams as 'hackers' to test the security of information stored on computers or transmitted through networks; the military uses red teams to address and anticipate enemy courses of action" (Ambrose and Ahern 2008, p. 136). Red teaming challenges the assumptions, generalisations, pictures or images that influence how we understand the world and how we take action, i.e. our mental models (Senge 1990). Dekker (2011, p. 39) argues that Murphy's Law is wrong: everything that can go wrong usually goes right, and then we draw the wrong conclusions. Red teaming can help us to see why, by pointing to alternative scenarios and outcomes. In the following, an adjusted risk assessment process is illustrated when using such a team to reveal surprises.

ADJUSTED RISK ASSESSMENT PROCESS USING A RED TEAM

The risk assessment process has four main stages, involving two analyst teams, referred to as teams I and II; see Figure 4.1. In **Stage 1** analyst team I performs a standard risk assessment, analyses risk and describes risk according to (A_1', C_1', Q_1, K_1). Here A_1' and C_1' are the specific events and consequences identified in the analysis, Q_1 is the assignments based on a

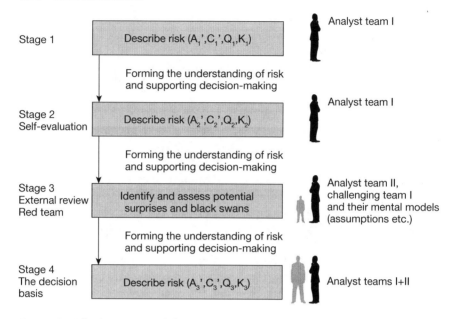

Figure 4.1 The four stages of the two-team risk assessment approach: A': specified events, C': specified consequences, Q: assigned probabilities (more generally assignments based on a measure of uncertainty/degree of belief), K: background knowledge for A', C' and Q (based on Veland and Aven 2014)

measure of uncertainty/degree of belief, typically assigned probabilities, and K_1 is the background knowledge that A', C' and Q are based on: data, information, justified beliefs (models, probability models, expert judgements, assumptions).

In **Stage 2**, analyst team I performs a self-evaluation of (A_1',C_1',Q_1,K_1), with a focus on the rationale for (A_1',C_1',Q_1,K_1) and highlighting the strength of knowledge of K_1 – an example of a key issue being the assumption deviation risk as discussed in Section 4.2. The updated risk description is denoted (A_2',C_2',Q_2,K_2). In many cases there would be no difference in the A', C' and Q compared to Stage 1, but the background knowledge K always changes as this review is performed, adding quality control of the various elements of the analysis process and a special judgement of the strength of knowledge of K_1.

In **Stage 3**, analyst team II challenges team I and their mental models (assumptions, etc.), acting as a red team (the devil's advocate). For example, they might:

- Argue for the occurrence of events with negligible assigned probabilities.
- Search for unknown knowns.
- Check how signals and warnings have been reflected.

A main purpose of this stage is to identify and assess potential surprises and black swans.

In the final **Stage 4**, the two analyst teams provide a joint risk description (A_3',C_3',Q_3,K_3), reflecting the input from both teams. The risk description provides a basis for understanding risk and supporting decision-making. See Veland and Aven (2014) for a practical guide and an example of the implementation of this process.

The analysis process may be rather resource heavy, but the full process should only be used in selected situations when the criticalities are considered high. In practice, simplified schemes may be developed to make the process feasible.

4.2 Semi-quantitative analysis

We consider here two examples at different levels (national and firm): a national risk level assessment, and a specific analysis concerning a LNG plant.

4.2.1 National risk level

In Norway national risk assessments (NRAs) have recently been conducted, the purpose being to provide a common and unified foundation for social safety planning across sectors and professions. The motivation for the NRAs was summarised in a recently published report on innovation in country risk management by the Organisation for European Cooperation and Development (OECD 2009):

> Central governments in particular have had to adopt a broader view on risk; one that is organised to address multiple hazards and vulnerabilities, and seeks to understand their interconnections rather than addressing each hazard and consequence separately. Implementing a broader view on risk requires the mobilisation and coordination of expertise from various government bodies and the private sector to increase breadth and depth of risk analysis for the purpose of better prioritising resource allocation.

The Norwegian NRA methodology is to a large extent inspired by methodologies developed in other European countries, primarily in the United Kingdom and the Netherlands (Veland *et al.* 2013). The Norwegian NRA process consists of four steps: 1) establish societal values; 2) identify hazards and threats; 3) conduct risk analysis; and 4) establish a common risk matrix.

The societal values are used for characterising the consequences of the identified hazards/threats. The following main categories of values are used (with associated consequence types to be used in the risk assessment in parentheses): life and health (loss of life, injury and disease, and physical strain); nature and environment (long-term damage to nature and the environment); economy (financial and material loss); social stability (social unrest and disturbance in daily life); and national governance and territorial control (reduced national governance and reduced control over territory).

Next, the identified hazards and threats are assessed with respect to risk. The assessment is based on the identification of a set of scenarios, referred to as plausible worst-case scenarios. From these, risk is described by predicting the consequences and assigning the associated probabilities of these scenarios. Each consequence type is given a score between A (very low) and E (very high); this is transformed to a numerical score, and the assigned scores for the nine consequence types of a given scenario are aggregated into one overall consequence score. The extreme case is a maximum score on each type, giving $1/9 + \ldots 1/9 = 1.0$, which is the maximum consequence score possible. If only two of the consequence types are relevant, the maximum consequence would be 2/9 and, depending on the score for these two types, we can get a total score between 0 and 2/9. The result is thus one overall score representing the entire spectrum of assessed consequences for one specific scenario.

The probabilities are assessed on the basis of historical data and expert judgements. The probability of risk events from intentional acts is assessed by considering the threat level posed by the capacity of malicious actor groups and the vulnerability of the targets defined in the scenario.

From these assessments, a standard risk matrix is established as a tool for presenting the overall results, reflecting the assigned probabilities of the hazards/threats/scenarios and the expected total consequences given the occurrences of these events.

The Norwegian Directorate for Civil Protection and Emergency Planning (DCPEP), which runs these NRAs, is in the process of adjusting the analysis approach to provide a more nuanced risk picture, which also reflects uncertainties. Up to recently, the analysis has to a large extent been based on a traditional (C,P) risk perspective as described in Section 2.2. Now a broader risk perspective seems to be adopted in line with ideas in this book (DCPEP 2013), but we will leave the DCPEP here and address the issue from a more general point of view. Given the general challenges raised in this type of risk assessment, how should we describe risk if we are to implement the new and broader risk perspectives described in Section 2.3? How should we reflect the knowledge (or lack thereof) and the surprise dimensions concretely when adopting these perspectives and not only the probability-based thinking and the use of standard risk matrices? This is the issue we will address in the following.

RISK ASSESSMENT

Let us first consider the national risk assessment challenge: how to describe risk in this case. The identification of risk events is the natural starting point, as for most types of risk assessment. To illustrate, let us focus on two such events (we refer to these as events A): storm, and terrorist attack. The probabilities of these events are assigned, and expected consequences are determined for each consequence type used and aggregated as explained above. We may find it adequate to report the results for specific consequence types and for all together. We obviously need to be careful in defining the events. For instance, a storm may be defined in relation to the Beaufort scale, and it could, for example, be reasonable to distinguish between a storm (Beaufort number 10: 24.5–28.4 m/s) and a violent or worse storm (Beaufort number 11 or higher: 28.5 m/s and over). Let us focus on the latter case. A probability can be assigned for this event to occur in Norway in the next year, based on historical records. Say that this probability is equal to p_0. This probability is based on specific knowledge (to a large extent formulated as assumptions), in particular that the historical data are representative of the future.

The expected consequences (we refer to the consequences as C) are then assessed, given the occurrence of the violent or worse storm, for each of the consequence types defined in Section 2.1 – for example, loss of lives. The expected value given to such an event is to be computed, i.e. probabilities times loss summed over all possible loss values, and it is obvious that in order to make reasonable judgements we need to condition on a number of possible situations, for example, where the storm occurs and when (e.g. day or night). The consequences could be strongly dependent on these situations, and then we also need to reflect on the relative probabilities that each of these situations will occur. Again we need to base our assessment on some knowledge (assumptions) to produce the probabilities. The different situations will produce consequences of different severity, and the standard risk matrix approach is to compute conditional expected values given the occurrence of the risk event. However, such an approach is obviously unfortunate as the variations in the consequences of the many situations covered are not reflected.

UNCERTAINTY INTERVAL FOR THE CONSEQUENCES

The natural way to cope with this problem is to develop an uncertainty interval for the unknown consequences of the event. To illustrate, let us first consider the case of one type of consequence only, the number of lives lost. For example, considering a number of situations, we may end up with [0,100] as a 90% uncertainty interval for the number of fatalities given the violent or worse storm. Hence, the analyst team has assigned a 90% probability for this number being in this interval. This number is far more informative than using only one estimate (say 2.3).

The uncertainty interval for the consequences represents the first extension of the standard risk description; see Figure 4.2. In theory we could also try to establish a probability distribution for the consequences given the risk event, based on the different situations, but this would be difficult as the situations covered would typically not be complete and the distribution would be seen as rather arbitrary. An uncertainty interval would be more robust in the sense that it is less sensitive to the choice of situations selected in the risk assessment.

However, an uncertainty interval and a distribution also have to be seen in relation to the assumptions made. The uncertainty interval and distribution clearly reflect variation – if a number of violent or worse storms occur, we will see them in different places and at different times. This is reflected by the interval and distribution, but judgements based on the analysts' knowledge are also a part of the basis for the established interval and distribution, for example when assessing the number of fatalities in a specific situation.

STRENGTH OF KNOWLEDGE

The uncertainty interval produced, [0,100] in the above example, does not express the strength of knowledge that supports it. Information about this strength would obviously inform the decision-makers and other stakeholders who will use the results of the risk assessment. The interval [0,100] is based on quite strong knowledge in this case, but the analysis could have been carried out quickly and based on poor knowledge and still obtain the same interval. The question is then how we should inform the decision-maker and communicate regarding this strength. What does it mean that the knowledge is strong or poor?

Intuitively, strong knowledge means a small or low degree of uncertainty and poor knowledge means a large or high level of uncertainty, but we have to be careful when referring to the uncertainty term here because it is not obvious what we are uncertain about. The concept of 'strength of knowledge' is considered more precise in reflecting the ideas that we would like to reflect.

Different rationales and implementation procedures for describing the level of the knowledge can be used. Two approaches will be presented here.

METHOD 1 TO ASSESS THE STRENGTH OF KNOWLEDGE

The first approach is based on a crude direct grading of the strength of knowledge that supports the probabilistic analysis, in line with the scoring used by Flage and Aven (2009). The knowledge is weak if one or more of these conditions are true:

(a) The assumptions made represent strong simplifications.
(b) Data/information are non-existent or highly unreliable/irrelevant.
(c) There is strong disagreement among experts.
(d) The phenomena involved are poorly understood, models are non-existent or known/believed to give poor predictions

If, on the other hand, all (whenever they are relevant) of the following conditions are met, the knowledge is considered strong:

- The assumptions made are seen as very reasonable.
- Large amount of reliable and relevant data/information are available.
- There is broad agreement among experts.
- The phenomena involved are well understood; the models used are known to give predictions with the required accuracy

Cases in between are classified as having a medium strength of knowledge.

METHOD 2 TO ASSESS THE STRENGTH OF KNOWLEDGE

The second approach is based on an identification of all the main assumptions on which the probabilistic analysis is based. These assumptions are converted to a set of uncertainty factors. For instance, in the example above, a main assumption for the probabilistic analysis was that the historical data are representative of the future. The corresponding uncertainty factor is the degree to which the historical data are representative of the future. The idea now is to perform a crude risk assessment of the deviations from the conditions/states defined by the assumptions. The aim of the assessment is to assign a risk score for each deviation, which reflects risk related to the magnitude of the deviation and its implications on the occurrences of the events A' and their consequences C'. This 'assumption deviation risk' score, which is to be seen as a measure of the criticality or importance of the assumption, captures the basic components of the risk description of the new risk perspectives, i.e. here; (i) the deviation from the assumptions made with associated consequences; (ii) a measure of uncertainty of this deviation and the consequences; and (iii) the knowledge on which these are based.

Below we indicate some practical ways of carrying out this crude risk assessment. A quick and very rough method would be to simply focus on the strength of knowledge scores (a) to (d) as above. For the example considered here (the historical data are representative for the future), we assign a weak strength of knowledge because several of the criteria (a) to (d) are considered true, and hence we consider the assumption deviation risk to be high. Similarly, we would have ended up with a low assumption deviation risk if a strong strength of knowledge score had been assigned.

A more detailed analysis can be conducted according to the following ideas.

Select one or more potential deviation events. In our case we may use three: the violent or worse storm occurrence rate for the relevant period of time is increased by a factor of 2, 10 or 100 compared to the one assumed. Then make a crude assessment of risk by considering the:

- magnitude of the deviation
- probability that this magnitude will occur

- the effect of this change on the consequences,

using score categories high, medium or low. We see that risk is basically expressed in line with the risk triplet of Kaplan and Garrick (1981); see Section 2.2.2. In the example case, the assessment gives a high risk score in the first scenario with an occurrence rate of 2, a medium score for the second scenario and a low score for the third, as the probability assigned is quite high for the first case (50%) but considerably lower for the others (1% and negligible, respectively); see Table 4.2. Next, an overall direct judgement is made of the strength of knowledge for the triplet risk assignments (the assumption considered is the same), again using the strong, medium and weak categories. In the case that a weak or medium score is assigned, the risk score based on the triplet assignment can be moved up one category, from medium to high risk, or from low to medium risk. In the example, a medium strength of knowledge score was assigned, but as the risk score was already high based on the risk triplet assignments, this led to no change in the overall risk score for this assumption.

More formally, the method can be stated as follows, using the general risk terminology (C',Q,K) and letting D denote the deviation. Hence the deviation risk can be expressed as $(\Delta C', Q, K_D)$, where $\Delta C'$ is the change in the consequences (which includes D), and K_D is the knowledge on which the $\Delta C'$ and Q are based. In the practical procedure outlined above, a score is first based on a judgement of D and related probabilities P (i.e. Q = P) (bearing in mind the implications for C'), and adjusted by a consideration of the strength of the knowledge K_D.

Based on these judgements we are ready to make conclusions about the overall strength of the knowledge that supports the probabilistic analysis:if we have a low number of assumptions with a high criticality/risk score, we would classify the strength of knowledge as high. However, if there are many assumptions with high criticality/risk scores, we would conclude that the strength of knowledge is poor. Furthermore, we may use a medium category or medium categories to reflect situations between these two.

Table 4.2 Risk scores for deviations from the assumption made in the risk assessment: the deviations are defined by the occurrence rate of the violent or worse storm scenario being a factor of 2, 10 or 100 higher than the assumed value (based on Aven 2013d)

Assigned risk score based on probability and consequences of deviation	High	x		
	Medium		x	
	Low			x
		2	10	100
	Deviation magnitude			

SOME REMARKS

In this example we referred to an occurrence rate for the violent or worse storm. If we had not introduced such a rate (the number of events is rare) we could not have defined a scale for the deviation, and we would not have been able to assess probabilities of the deviations. Then we could have restricted our attention to the simple (a) to (d) approach, or we could have looked for alternative ways of defining deviations from the assumptions. One such approach would simply be to say that we consider the deviations, defined by stating that the probability of occurrences of violent or worse storm scenarios to be a factor of 2, 10 or 100, are higher than those used in the risk assessment. Then there is no uncertainty in the probability assignment (because it expresses the degree of belief of the assigner), and the risk score has to be based on judgements about the implications for the consequences C and in particular the strength of knowledge supporting the various scenario probabilities.

The above analysis is linked to the uncertainty interval of the consequences, C, but is equally applicable to the probability assignment of the risk event, P(A). Hence, we can establish a strength of knowledge score for both the uncertainty interval and the P(A). To simplify and avoid too many score attributes, we may reduce these to one – for example using the one with the lowest score, i.e. closest to poor. This thinking has been implemented in Figures 4.2 and 4.3.

Looking at Figure 4.2, one may think that there should also be an uncertainty interval for the probability P(A) of the risk event A, the violent or worse storm, but the two axes are different. The probability assigned is a knowledge-based (judgemental, subjective) probability, meaning that it is conditional on the available knowledge K. As before we write $P(A) = P(A|K)$ to show this dependency. This probability reflects the assigner's degree of belief based on his/her knowledge. There is no uncertainty in the assigned value, as this would presume that there was a correct probability value. However, the values assigned are dependent on the knowledge available, and as discussed for the uncertainty interval above, the strength of the knowledge is also of interest as it reflects in some way the 'quality' or 'goodness' of the probabilities assigned.

Having rejected the idea of uncertainties in the assigned knowledge-based probabilities, it is necessary to add that there could sometimes be an imprecision problem related to assigning such probabilities. When we assign a probability, say the value 0.2, we have in fact established an imprecision interval [0.15, 0.24] or [0.150, 0.249], because all numbers in this interval are equal to 0.2 when working to only one significant figure. It is possible to develop a theory of uncertainty representations based on such intervals, but this will not be further discussed here; we refer to the discussions in Appendix A and Aven and Zio (2011).

Typically, we will not have more than one occurrence of a specific risk event, as they are extreme events. However, if we are considering risk events with many occurrences during the period studied, we need to replace the probability in Figures 4.2 and 4.3 with a number of events. Then the two axes become similar in the sense that uncertainty intervals are relevant for both. We can present an expected number of events occurring, and then it is clear that we need to consider the uncertainties in this number, because the expected value can produce poor predictions of this number.

Diagrams like Figure 4.2 and Figure 4.3 are often shown in matrix form with the consequence dimension either reflecting expected consequences given the event, E[C|A], or a specific consequence, say one fatality. In the latter case we then have no uncertainty in the consequences because the consequences are specified by the scenario: an event occurs that lead to one fatality. In this case we obtain a risk matrix as shown in Figure 4.4, where the c_is are the specific consequences that the events can lead to. Hence, one score in the matrix expresses the probability that a scenario occurs with this particular consequence. In the figure three possible categories for the probability are used (these could be replaced by intervals).

A simplified version of Method 2 can be carried out by using scores such as (a) to (d) defined above for each assumption, and then integrating this assessment with an analysis of the sensitivity of the results of the probabilistic result with respect to changes of the assumption (uncertainty factor), see Flage and Aven (2009).

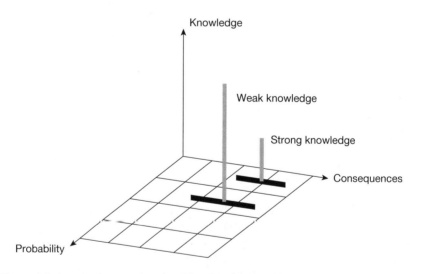

Figure 4.2 A way of presenting the risk related to a risk event when incorporating the knowledge dimension (based on Aven 2013d)

Probability
Consequence, impact
Poor background knowledge
Medium background knowledge
Strong background knowledge

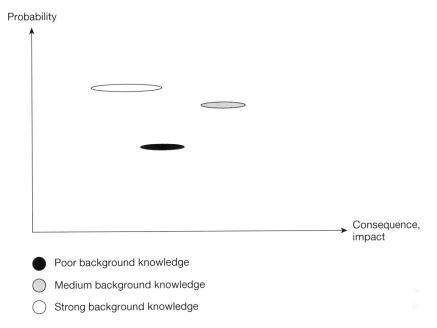

Figure 4.3 An adjustment of the risk presentation shown in Figure 4.2 (based on Veland and Aven 2014)

Probability
p_1
p_2
p_3
c_1
c_2
c_3
Consequences
Poor background knowledge
Medium background knowledge
Strong background knowledge

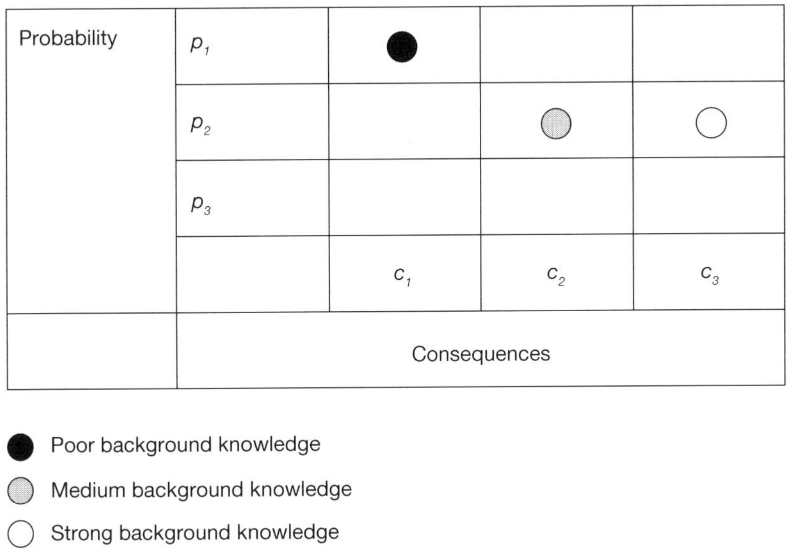

Figure 4.4 A risk matrix based on specified consequences and reflecting the strength of knowledge (based on Veland and Aven 2014)

BLACK SWANS

Next, we would like to address black swan type events. A procedure such as that described in Section 4.1.2 is adopted. Here is an illustration of parts of the analysis.

A list of all types of risk events having low risk with regard to the three dimensions, assigned probability, consequences, and strength of knowledge, is produced. An example of such an event could be a storm with a score 10 on the Beaufort scale that kills a large number of people, or a specific type of terrorist attack. These events would not be shown on the national risk maps because they are considered to have low risk.

Secondly, a review of all possible arguments and evidence for the occurrence of these events is provided, for example by pointing to historical events and experts' judgements that are not in line with common beliefs. To carry out these assessments, there is a need for a different group of people from those performing the risk assessment. Of course, the idea is to allow for and stimulate different perspectives, in order to break free from common beliefs and obtain creative processes.

This list of potential black swan type events, with associated risk descriptions reflecting this type of argument and evidence, are reported along with the risk events having the highest risk scores according to assigned probability, consequences, and strength of knowledge.

The above approach for assessing a risk event is applicable to a storm as well as to other events, including intentional acts (such as terrorist attacks). However, the knowledge dimension is much more dynamic in the case of intentional acts. Events in other places in the world and the results of surveillance and intelligence work could quickly change the risk assessment of such acts, both in terms of the probabilities as well as the strength of knowledge.

Think of the killing in Norway on 22 July 2011, when a man placed a car bomb outside the government office and then massacred a number of people on the island of Utøya outside Oslo. This event came as a surprise to the police security services, relative to their knowledge. It was thus a black swan for them. Hypothetically, if the police security services had carried out a risk assessment before 22 July 2011, they could have produced a low risk for such a type of event according to the assigned probability, consequences, and strength of knowledge. Given the above recommended additional procedure, however, the event would have been included in the list of black swan events of category II, with arguments and evidence provided for its occurrence. Of course, this approach would not have ensured that suitable actions would have been taken to stop this attack; the point here is limited to a consideration of the way we perform the risk assessments and describe the risk.

4.2.2 Risk assessment of a LNG plant

An LNG (Liquefied Natural Gas) plant is planned and the operator would like to locate it only a few hundred metres from a residential area (Vinnem 2010). Several quantitative risk assessments (QRAs) are performed in order to demonstrate that the risk is acceptable according to some pre-defined risk acceptance criteria. In the QRAs the risk is expressed using computed probabilities and expected values.

However, it turns out that the assessments and the associated risk management meet with strong criticism. The neighbours and many independent experts find the risk characterisation insufficient – they argue that risk has been reported according to a too narrow risk perspective.

The risk assessments carried out in this case were all based on a traditional (C,P) risk perspective as outlined at the beginning of Section 2.2.2. Now, how should we describe risk if we have instead adopted a broader risk perspective in line with the (C,U) risk perspective?

The risk assessment for this plant is similar to the one reported in the previous section and, to avoid repetition, only a few further comments will be provided here. Risk is also described in this case through risk matrices, and the above analysis then applies. However, other risk metrics are also used: individual risk and f-n curves. The f-n curves show the assigned probability for accidents occurring with at least x fatalities as a function of x (Bedford and Cooke 2001). To these curves, as in the previous section, we add a dimension reflecting the strength of knowledge on which these assignments are based. There is no uncertainty interval here as in the risk matrix in Section 4.2.1. The same is the case for the individual risk, which expresses the assessor's probability that an arbitrary but specific person will be killed during a specific year. As for the f-n curve, we need to add the strength of knowledge component. Black swans are analysed as in the previous section.

For this particular case, the strength of knowledge is assessed by reviewing the list of assumptions and assessing the importance of these according to the procedure outlined in the previous section. To compute the risk metrics, for example, a number of assumptions (uncertainty factors) were made (Aven 2011a):

(1) The event tree model.
(2) A specific number of exposed people.
(3) A specific fraction of fatalities in different scenarios.
(4) The probabilities and frequencies of leakages were based on a database for offshore hydrocarbon releases.
(5) All vessels and piping are protected by monitors, hydrants, etc.
(6) In the event of the impact of a passing vessel on a LNG tanker loading at the quay, the gas release would be ignited immediately (by sparks generated by the collision itself).

Assumption deviation risk scoring of these and many other assumptions (uncertainty factors) were assigned; see Table 4.3 (only the scores of these six assumptions are shown). First, an assessment was carried out using the criteria (a) to (d) referred to in the previous section, and then the more detailed method was applied where feasible. Let us look into one of these assumptions – the last one.

Assumption (6), among others, was given the highest risk score. Several experts argued against this assumption. One of these writes:

> The implication of this assumption was that it was unnecessary to consider in the studies any spreading of the gas cloud due to wind and heating of the liquefied gas, with obvious consequences for the scenarios the public might be exposed to. Such a very critical assumption should at least have been subjected to a sensitivity study in order to illustrate how changes in the assumption would affect the results, and the robustness of the assumption discussed. None of this, however, has been provided in any of the studies.
>
> (Vinnem 2010)

An uncertainty factor related to assumption (6) can be formulated as the time duration between the gas release and the ignition for the relevant scenario. In the assumption deviation risk assessment we question how the consequences C are influenced by increasing this duration from zero to (say) one hour, and how likely it is to have such a deviation. A high score was assigned for this assumption as both the deviation probability and the consequence of the deviation were considered to be quite large. The strength of knowledge judgements of the triplet assignments were considered to be moderately large, and this further supported the high risk score for this assumption.

Due to the many assumptions giving a rather high risk/criticality score, the overall conclusion was that the strength of knowledge of the analysis needs to be classified as weak or at best medium. It would be essential for this analysis and conclusion to accompany the numerical results produced. Obviously, care has to be shown when comparing the produced f-n curves and the individual risk numbers with pre-defined risk acceptance criteria, without also reflecting on the strong dependencies of the assumptions made. We refer to Section 5.1.2.

Table 4.3 Assumption deviation risk scoring of six assumptions made in the risk assessment of the LNG plant (based on Aven 2013d)

High risk		x	x			x
Moderate risk	x					
Low risk				x	x	
Assumption	1	2	3	4	5	6

This risk/criticality scoring can also be used as a screening procedure for where to place the focus to improve the risk assessment. The assumptions with the high score should be examined to see whether they can be dealt with in some way and removed from the highest risk/criticality category. However, in practice it is never possible to carry out a quantitative risk assessment without making many assumptions.

4.3 Using signals and warnings

A challenge for risk management is to avoid missing or ignoring early signals and precursors of serious events, or, on the other hand, exaggerating them. It is common practice to refer to false negatives (no indication of a risk situation when one is actually present) and false positives (erroneous signals indicating some risk situation is present when it is not), but how can we make judgements about these 'errors' when we do not know the outcomes of the events or situations under observation before they occur? It is easy to identify (and claim) that we missed a risk event or situation with hindsight a posteriori, when the accident, disaster or crisis has occurred, but how can we know in advance that we are missing, ignoring or exaggerating signals or precursors, given that we are typically exposed to a large number of threats/ hazards? The reference for our evaluation of the signals and precursors cannot be the unknown consequences or outcomes of events yet to occur. The only possible way out seems to be to rely on the results of risk and uncertainty assessments, where the warning system itself can be viewed as a form of risk assessment. However, risk assessment has its limitations as a tool for this purpose, and in cases when the knowledge base is not strong, we need to base the judgements on hypotheses and assumptions, and we may act too slowly (or too quickly). An example is the AIDS epidemic which was detected in the United States by the Center for Disease Control in 1981, and given that it had probably been spreading for decades, the response was rather slow (Paté-Cornell 2012).

Let us look a little deeper into this issue. The most common approach for this purpose is simply statistical analysis. To illustrate the analysis, consider a health example. The question is whether John suffers from a specific disease.

Here the standard set-up is the introduction of a probability model, with an unknown parameter r. If r is in a set R_1 the person is considered sick, if r is not in this set, he is not. Some data X are observed (signals and warnings), and the hypothesis that John is sick is tested. If the data X are sufficiently 'extreme', it is concluded that r is in the set R_1, i.e. he is sick. We cannot know for sure if this is really the case, but the method is constructed such that the frequentist probabilities for errors are small. Alternatively, a Bayesian approach can be used, where the posterior (subjective) probability that John is sick (r is in the set R_1) is computed using Bayes' formula, on the basis of the presumed probability model, a prior distribution of r and the observations X. We omit the details here as this is general statistical theory found in any text book in statistics (see e.g. Meeker and Escobar 1998).

False negatives (no indication of a risk situation when one is actually present) mean in this example that John is sick but the signals do not indicate this, whereas false positives (erroneous signals indicating some risk situation is present when it is not) occur when the signals show that John is sick when he is not. The statistical approaches as indicated above, whether frequentist or Bayesian, are well established for treating the errors related to false conclusions within their frameworks.

In the frequentist approach we are able to control the frequentist probability where we conclude that the null hypothesis is wrong when this is not in fact the case. In our example, if we define the null hypothesis as John being sick, the test is constructed such that there is a low frequentist probability that we conclude that John is not sick when in fact he is sick. In the Bayesian case, we simply produce the probability of John being sick given the present knowledge as a basis for the decision-making.

Both approaches are built on probability models and risk is typically understood as (A',C',P_f), where A' and C' are the specific events and consequences identified in the analysis and P_f is frequentist probability (chance). As discussed in Sections 2.2 and 3.1, the framework and risk perspective makes it difficult for us to break free from this model. Risk per se is defined through chance, and consequently there is no 'acceptance' for rejecting this model. Given new observations, in the form of new data, a reinterpretation of the knowledge basis might be needed. This applies in particular to the probability models adopted. The data could be surprising relative to the models used, and new insights about the phenomena studied could challenge basic assumptions of the analysis.

An extended approach is thus required, and a framework for such an approach is conceptualistic pragmatism, which links probabilistic analysis with knowledge theory and the quality movement with a focus on continuous improvement (Bergman 2009). Aven and Bergman (2012) discuss how this conceptualistic pragmatism relates to risk assessment. However, we shall go no further into this approach here; the key idea is that we need to rethink the framework in view of new observations and knowledge. For example, in the John scenario, the observations (signals) may be of a form that requires a modification of the probability model adopted, or even a new model.

Of course, good analysts should always explore the robustness of the assumptions made and the probability models used (Bergman [2009] refers to discussions by Lindley [1985] related to the so-called Cromwell's rule), but in practice such reinterpretation of the knowledge basis is seldom carried out unless the framework adopted allows for and encourages such reinterpretations. If Risk = (C,U), reflection on the model is easier to incorporate. Following this approach, chances are not automatically introduced; they have to be justified. In contrast, the relative frequency-based perspectives (A,C,P$_f$) are founded on the existence of probability models and chances. This makes the (A,C,P$_f$) perspectives less open for this type of reflection. It is more difficult to escape from the initial assumptions. The implications of this are not only of theoretical interest; they strongly affect judgements about risk acceptance and risk management in general. See the discussion in Section 5.1.2.

There are many other possible analysis approaches to assess and treat the signals and warnings. We shall look briefly at two of these – adaptive risk analysis and robust analysis – which are of special interest for our setting.

ADAPTIVE RISK ANALYSIS

This type of analysis is based on the acknowledgement that there is no one best decision; rather, a set of alternatives should be dynamically tracked to gain information and knowledge about the effects of different courses of action. One of the challenges facing risk analysts and managers today is that systems and activities are often characterised by large/deep uncertainty. Examples include analysing and managing complex financial derivatives, novel technology, preparations for climate change, emerging diseases, and complex, highly intertwined infrastructure. The uncertainty regards potential future events and consequences, and, in the case of deep uncertainty, an essential feature is the lack of justifiable prediction models.

In response to situations with deep uncertainties, robust and adaptive risk analysis is often recommended; see for instance Lempert *et al.* (2004), Kasperson (2008), Walker *et al.* (2003) and Cox (2012). Here we look first at adaptive risk analysis. On an overarching level, the basic process is straightforward: one chooses an action based on broad considerations of risk and other aspects, monitors the effect, and adjusts the action based on the monitored results (Linkov *et al.* 2006). A central idea is that since uncertainty is pervasive, one optimal management choice is not achievable; rather, we have a range of often competing decision alternatives, which we track dynamically to gain information and knowledge about the system and about the effects of different courses of action.

Current applications all seem to be based on rather traditional perspectives on risk, using probability to represent uncertainties (e.g. Wintle and Lindenmayer 2008). However, such risk perspectives are not especially suited for an adaptive risk management context, where the importance of knowledge and uncertainties is so critical. We adopt a broader risk perspective in line with the (C,U) model to better reflect these aspects.

To illustrate the approach, let us return to the talk example (Section 1.2). John is to deliver a talk at a scientific conference. His long-term goal is to give brilliant talks. He is now in the middle of his talk. He has a good feeling and is getting positive feedback from the audience. Then he observes that a person, Ralph, who is well-known for cavilling behaviour, is among the audience. John is not prepared for this. He has worked hard to be prepared for the talk and has even thought about possible surprises, but he has not seen this situation coming. He now faces considerable risk, and the uncertainties are large. If he does not take appropriate action, he infers that the outcome of his talk could be far from successful. He thinks quickly through the alternatives:

(1) Act as if nothing has happened and hope for the best.
(2) Try to get some positive interaction with Ralph.

In line with the collective mindfulness criterion to be 'sensitive to operations', he chooses the second alternative. John looks at Ralph and makes a remark intended to give Ralph some recognition for being among the audience. The response is immediate and Ralph returns a smile which John interprets as a positive signal. Given the response, John judges the probability that Ralph will initiate some cavilling behaviour to be small, less than 10%. However, he sees that the strength of knowledge supporting this judgement is poor, and hence the risk is still found to be very large. John shifts for a period to decision alternative (1), trying to forget Ralph and the associated risk. After some time, John cannot avoid seeing that Ralph's facial expression has changed and John's assessment is that the risk and uncertainties have now reached a level that requires urgent action. He judges the probability of a 'disaster' is very high – at least 50% if he does nothing. The strength of knowledge is not an issue, as the large probability number and the large uncertainties require a cautionary reaction. He decides to turn to Ralph and give him positive attention in the form of some words that offer recognition of his work, as an attempt to get in first. The result is overwhelming; Ralph smiles and his facial expression changes dramatically. The rest of the talk is carried out with no further episodes linked to Ralph.

Following the response from Ralph, John described risk by a probability (actually an interval probability) and the strength of knowledge. Alternatively, he could have just made a qualitative judgement of the combination of consequences and associated uncertainties. In either case he would have come to the same conclusion: the risk was very high.

We see that restricting the risk considerations to probability judgements could mislead John in his decision-making. The knowledge and uncertainty dimensions, which are important for the decision-making, are then not sufficiently reflected.

We see from the example how well the collective mindfulness principle to be 'sensitive to operations' matches the risk understanding and treatment.

For an analogous example linked to an offshore oil and gas operation, see Bjerga and Aven (2014).

Bayesian decision analysis provides a strong theoretical framework for choosing optimal decisions in the case of information in the form of signals and warning, but it is difficult to use in practice in many cases. Instead, we may search for procedures that prescribe what to do in situations with given signal/warning levels. The idea is simply to make such an assessment for different signal/warning levels and establish an adequate decision rule for how to act in the different cases. This would give a level of preparedness in the case of specific signals/warnings, but would not necessarily provide much support in the case of surprising events. It would be impossible to prescribe what to do in all cases; hence the approach needs to be supplemented with other methods.

There is a common belief among many engineers and managers that to manage an activity and avoid accidents and perform operations as planned, it is sufficient to develop procedures and ensure compliance with these. In practice, however, such a compliance perspective fails for non-trivial activities because a perfect system cannot be developed; surprises always occur. The system understanding is too static, and improvements and excellence are not sufficiently stimulated. We will discuss this issue in more detail in Section 5.2.

We have to understand that to obtain excellence and avoid accidents we need to acknowledge the performance, risk and knowledge 'dynamics'. We need to see beyond compliance. For many types of systems, the signals and warnings are of a form that requires judgements and actions that need considerations beyond specified procedures. Think of the example of John's talk. If John specifies in advance exactly what to do in a set of situations, and rehearses these a lot, he will clearly be able to deal with the occurrence of a situation that is in the planning set. However, surprises will always occur and then he needs to be able to absorb and analyse relevant information and take adequate actions. Adaptive risk analysis can be a useful tool in this respect, and so can abductive thinking:

> You observe a fact... In order to explain and understand this, you cast about in your mind for some glimmering theory, explanation, flash and so forth. The process of abduction takes place between the result (observed fact) and the rule (explanation), and concludes with the positioning of a hopefully satisfactory hypothesis.
>
> (Harrowitz 1983, p. 183)

As presented by Pettersen (2013), abduction can be seen as the process of noticing an anomaly and getting an explanatory hunch (Chiasson 2001). By means of abduction a new idea (or hypothesis) is brought up from the region where 'all things swim'. This can be shown as a three-step process:

(1) A surprising fact is noticed.
(2) An aesthetic (unfettered) exploration of qualities and relationships is made.
(3) Abductive reasoning is applied to make a guess that could explain the surprising fact. (Chiasson 2005)

Returning to our talk example, abduction means noticing that Ralph is among the audience, exploring why he is present and providing a hypothesis to explain it. For example, John may hypothesise that Ralph is in the audience to take revenge from an earlier event. Then testing can be carried out to prove or disprove the hypothesis. As we remember, John quickly looked at Ralph and made a remark intended to give Ralph some recognition for being among the audience. The feedback to some extent disproved the hypothesis and a new one arose: Ralph would not show cavilling behaviour. We see how the thinking fits nicely into the adaptive risk analysis performed above.

It is also in line with fundamental ideas of the quality discourse, which highlights that knowledge is built on theory (Lewis 1929; see also Bergman 2009). As formulated by Deming (2000, p. 102), rational prediction requires theory and builds knowledge through systematic revision and extension of theory based on comparison of prediction with observation. Without theory experience has no meaning, and without theory there is no learning. If we consider John's long-term development to become a brilliant speaker, he bases his work on the theory in this book, among other theories. He performs and compares the outcomes with the theory, and there will be a continuous improvement process (using the basic steps: plan, do, study and act), which may cover adjustment/developments of the theory and how to interpret it in practice. The author of the present book similarly believes in the theory here presented as a useful perspective and a good way of developing the proper understanding, assessment and management of risk. The theory is justified by the arguments given, and through future observations it can be adjusted and further developed. See also Sections 5.2 and 5.4.

ROBUST ANALYSIS

We can write the problem in general terms as (C,P,u,a), where C are the consequences of the actions, P the probabilities of C given the actions, u the utility function of C, and a the actions (in line with Cox 2012). However, it may be difficult to assign some of these values, for example P, when the uncertainties are large. A robust approach is then required. To carry out meaningful robustness analysis, it must be made clear which unknown quantities and parameters are to be considered for the analysis, and what the variation or uncertainty is reflecting. Only then can a distinction be made between what should be included in the robustness analysis and what is left for other types of considerations (for example, political ones). To formalise the challenge somewhat, let g be a criterion for which we would like to decide on a 'good' solution. An example of such a criterion is expected costs or expected utility. Theoretically we can write g as a function of C, P, u and a, leading to:

$$g = g(C,P,u,a)$$

The traditional analysis approach is based on specific values of P and u. A robust analysis is about finding actions a where the need for specifying C, P and u is relaxed in some way. However, there are a number of ways in which the robust analysis can be carried out, even for a case where the P is only allowed to be intervals instead of a specific value; see for example Goerigk and Schöbel (2011). In Aven and Hiriart (2013), a simple investment model based on expected costs is studied and several robustness schemes are considered relevant to reflect imprecision in the P values. This fact – that there are so many ways of looking at robustness, and none can be argued to be more natural and better than others – should point to the need for a cautious policy in arriving at conclusions about the best action a, with reference to one particular robustness scheme. Aven and Hiriart (2013) conclude that the robustness set-up presented for the precise formulation of this type of problem provides a useful platform for gaining insights into the need for safety investments. As there are many ways of defining a robust solution, alternative approaches should be applied to inform the decision-maker. A sensitivity analysis, showing the optimal investment levels for different parameter values, should also be presented, followed by a qualitative analysis of the arguments supporting the different parameter values.

All this underlines the necessity for seeing the robustness analyses as no more than decision support tools that need to be followed up with an evaluation (with reference to managerial review and judgement) which also reflects on:

(1) The choice of criterion g. In practice the criterion always has some limitations. For example, if expected values are used, aspects of risk aversion may be ignored; see Section 5.1.1.
(2) The assumptions on which the analyses are based and the possibility of black swans. Although the robustness scheme allows for fewer assumptions than if specific numbers are used, any analysis needs to make assumptions to be able to carry out the calculations and considerations. For example, it may be assumed with certainty that a quantity lies in the interval [a,b], but the true value could turn out to be outside. Another assumption could be the use of a specific probability model. The result of the analysis needs to be seen in relation to these assumptions. The surprise dimension cannot be ignored by the decision-maker and needs to be given due attention in the managerial review and judgement. Requesting the analysis group to perform a type of surprise assessment would not solve the problem, as the point is that the decision-maker has to consider the possibility of surprises that extend beyond the knowledge and apparatus of the chosen analyst group.

The robustness analyses can provide insights and decision support, and the use of different types of such analyses may be useful to inform the decision-maker in many cases, but we have to acknowledge that there is often considerable arbitrariness in the choices made by the analysts, and all the tools used have significant limitations. As a consequence, care should be taken not to draw strong conclusions based on the results of such analyses. There is always a need for a managerial review and judgement, which places the results of the formal analyses in a broader context where the limitations and boundaries of the analyses are taken into account before a decision is made. A decision that is determined mechanically by the analytical approach can seldom be justified. We will return to this issue in Chapter 5.

4.4 Methods for evaluating risk assessments and the understanding of risk

The ideas of risk presented in this book can be used to perform quality evaluations. Three examples will be presented here:

(i)   Evaluation of the risk assessment approach used in a job safety analysis.
(ii) Evaluation of a concrete risk assessment performed (a job safety analysis).
(iii) Evaluation of the understanding of risk among the persons carrying out the assessment.

In a job safety analysis, hazards linked to the various subtasks of the activity studied are identified, and risk is described by a risk matrix covering consequence and probability. Based on this risk description, an overall judgement about safety and risk acceptability is made. An evaluation of this method quickly reveals that the way risk is described is based on a probability-based view of risk, and, as discussed in Section 2.2.2, this perspective can be criticised as not properly reflecting the strength of knowledge that the assessments are based on, uncertainties, surprises and black swans. It is beyond the scope of the present book to provide a detailed solution to this challenge, but some ideas have been suggested above in Section 4.2, by incorporating judgements of the strength of knowledge on which the probability-based judgements are based, and making assessments of the so-called assumption deviation risk. Specific related black swan analysis should also be considered, e.g. some form of red-team analysis as mentioned in Section 4.1.2, using the four-stage approach presented there.

Then we look at the second example: a risk analysis team has conducted a risk assessment in relation to a critical maintenance operation, a job safety analysis. To evaluate the quality of this analysis, we apply the principles adopted in this book, presuming that the intention now is to conduct the analysis in line with the recommended principles. A check is made that the analysis covers what can go wrong, causes, barriers, consequences and likelihood, as well as issues linked to uncertainties, knowledge, surprises and black swans. A key point is that the analysis is clear on the key assumptions and beliefs on which the judgements are based, and that these assumptions and beliefs are challenged in an appropriate way. Key control questions relate to the data and information supporting the assessment and the ability of the analysts to provide a meaningful presentation of the results, which is clear on what the analysis provides and what the limitations are. In the concrete example referred to in Aven (2013g), it transpired that the results relied on a key assumption which was not questioned in the uncertainty analysis, despite the fact that critical questions had been raised in other contexts. However, these questions were not known to the analysis team that performed the initial risk analysis.

This critical maintenance operation is also considered in the third example. Here we evaluate the risk understanding of the personnel who are to carry out the work. The focus is on the potential hazards, their causes and consequences, including barriers, as well as the likelihood. Special attention is paid to signals and warnings that may give an indication of the occurrence of more severe hazardous situations, and how to detect these signals and warnings and make adequate adjustments. The degree to which the insights provided by the relevant risk assessments are known and understood is questioned. This relates to all the issues mentioned above (what can go wrong, causes, barriers, consequences and likelihood), but also covers issues linked to uncertainties, knowledge, surprises and black swans (if such issues have not been addressed by the risk assessment, a separate analysis of such issues should be added in line with the recommended thinking). A key issue is to reveal the main assumptions and beliefs on which the judgements are based, and how to cope with surprises relative to these assumptions and beliefs. In the specific case considered in Aven (2013g), the fourth mindfulness principle (commitment to resilience) was suggested as a weakness, since the issue of robustness and resilience had not been properly thought through. The current standard in the company for how to carry out such operations fails to stress the importance of commitment to resilience. Moreover, it was identified that key personnel had a lack of understanding of the system, and that important experience from recent operations had not been captured by the risk assessment and was not known to all the personnel involved in carrying out the critical operation. See Veland and Aven (2014) for some guidelines on how to carry out evaluations such as (ii) and (iii).

4.5 Remarks

From time to time I hear people, particularly risk analysts, question the need for improved risk assessment methods. They argue that the main challenge associated with more effective use of risk assessments in industry and the public sector is not the analysis methods used, but the use of the assessment in a decision-making context. Instead of using assessments actively as a tool to support decision-making about the choice and arrangement of risk-reducing measures, assessments are used for verification purposes, where the conclusion of acceptable risk according to some specified decision criteria (risk acceptance criteria) is trivial; see the discussion in Section 5.1.2.

The response to this argumentation is – yes, we do need to improve the way risk assessment is used in decision-making contexts, but we also need to continuously improve the tools we use to assess risk. These tools have been shown to be effective for many types of situations, but they are not sufficiently developed when it comes to two features:

(1) The knowledge dimension.
(2) The surprise dimension.

The present book provides appropriate concepts and an understanding of what these features actually mean. In addition, it also provides ideas and principles for how to better incorporate these features in our analyses.

In safety work we identify events and scenarios, for example hydrocarbon leaks, and there are barrier systems that combat these events if they do in fact occur. Many types of events happen during the course of a year, but they do not have serious effects because the barriers work as intended. However, in many cases the margins are small; with minor changes in scenarios, potential disasters could have become a reality. When a major accident occurs, it is often because there are many 'surprising events' occurring together. Just to have an understanding of what all this means – what terms such as surprising, unexpected, risk, etc. signify – is in itself important for the treatment of these issues. The book contributes to this understanding (item 1 above).

In addition, the book provides specific help as to how we should proceed in order to improve the understanding of risk and the assessment and management of risk where black swans are included. Knowledge and uncertainty are key concepts. Black swans are surprises in relation to someone's knowledge and beliefs. In the September 11 event some had knowledge, but others did not. In the Fukushima case it was the judgements and probabilities that were essential, but they were based on data, information and beliefs, so here too the issue was about knowledge. Hence we must think beyond current practice and theory. We need new principles and methods. This book contributes to such principles and presents examples of specific methods. It lays the foundation for the further development of a set of specific methods, to better assess and manage potential surprises and black swans.

When performing risk assessments, assumptions are required. In Section 4.2 we looked at one approach for assessing the importance of these assumptions. Several other approaches can be used. Here we will draw attention to the work by Funtowicz and Ravetz (1990) and Kloprogge *et al.* (2005, 2011) linked to the NUSAP system. NUSAP was proposed by

Funtowicz and Ravetz (1990, 1993) to provide an analysis and diagnosis of uncertainty in science for policy by performing a critical appraisal of the knowledge base behind the relevant scientific information. The basic idea is to qualify quantities using the five qualifiers of the NUSAP acronym: Numeral, Unit, Spread, Assessment, and Pedigree. Our focus here is on the last dimension, the pedigree, which resembles the criteria used in Section 4.2 to assess the strength of knowledge supporting the quantitative analysis.

The pedigree approach is founded on a set of criteria to trace the origin of certain assumptions (hence, a 'pedigree of knowledge') and qualify the potential value-laden character of these assumptions (Laes *et al.* 2011). The criteria used to discuss the assumptions are (van der Sluijs *et al.* 2005a, 2005b):

- *Influence of situational limitations*: the degree to which the choice of the assumption may be influenced by situational limitations, such as limited availability of data, money, time, software, tools, hardware, and human resources.
- *Plausibility*: the degree, mostly based on an (intuitive) assessment, to which the approximation created by the assumption is in accordance with 'reality'.
- *Choice space*: the degree to which alternatives were available to choose from when making the assumption.
- *Agreement among peers*: the degree to which the choice of peers is likely to coincide with the analyst's choice.
- *Agreement among stakeholders*: the degree to which the choice of stakeholders is likely to coincide with the analyst's choice.
- *Sensitivity to the view and interests of the analyst*: the degree to which the choice of assumption may be influenced, consciously or unconsciously, by the view and interests of the analyst making the assumption.
- *Influence on results*: in order to be able to pinpoint important value-laden assumptions in a calculation chain it is important not only to assess the potential value-laden-ness of the assumptions, but also to analyse the influence of the assessment on the outcomes of interest.

Each assumption is given a score between 0 and 4; see Table 4.4 for each of the criteria.

Table 4.4 Pedigree scheme used to assess assumptions (based on van der Sluijs *et al.*, 2005a, 2005b)

Score	Influence of situational limitations	Plausibility	Choice space	Agreement among peers	Agreement among stakeholders	Sensitivity to views of analyst	Influence on results
4	No such limitations	Very plausible	No alternatives available	Complete agreement	Complete agreement	Not sensitive	Little or no influence
3	Hardly influenced	Plausible	Very limited number of alternatives	High degree of agreement	High degree of agreement	Hardly sensitive	Local impact in the calculations
2	Moderately influenced	Acceptable	Small number of alternatives	Competing perspectives	Competing perspectives	Moderately sensitive	Important impact in a major step in the calculation
1	Importantly influenced	Hardly plausible	Average number of alternatives	Low degree of agreement	Low degree of agreement	Highly sensitive	Moderate impact on end result
0	Completely influenced	Fictive or speculative	Very ample choice of alternatives	Controversial	Controversial	Extremely sensitive	Important impact on end result

The pedigree methodology is used to identify, prioritise, analyse and discuss uncertainties in key assumptions. The pedigree assessment (based on Laes *et al.* 2011):

- Provides qualifying assumptions when they are communicated to various stakeholders.
- Evaluates how assumptions relate to different perspectives and frameworks, as held by involved actors.
- Generates suggestions for improvements in the calculation chain to deal with disagreements and divergence over assumptions as well as suggestions for communicating this type of risk assessment results.
- Conveys a more adequate image of such risk assessments.

There are many versions of the pedigree approach, but they are similar and can be used in the same way as indicated by the (a) to (d) method presented in Section 4.2.2. However, adaptation to the specific case under consideration is required. The general form of the pedigree as presented above has a typical societal safety/security scope and extends beyond the strength of knowledge considerations in Section 4.2 – for example, in that the views of external stakeholders are also incorporated. See Berner and Flage (2014) for a detailed discussion of the link between the (a) to (d) criteria and the pedigree approach.

Bibliographic Notes

There are a number of text books in risk assessment (e.g. Vose 2008, Zio 2007, Aven 2008). This chapter covers contributions to the field, specifically addressing the knowledge dimension and surprises, which are not found in these books. The key sources for this chapter are Aven (2013d) and Aven and Krohn (2014). In addition we have given considerable space to the anticipatory failure determination (AFD) method developed by Kaplan *et al.* (1999). This method is seen as an interesting approach for identifying potential failure scenarios and how to meet these, and is not yet commonly referred to in the risk assessment literature. We have addressed adaptive risk analysis, Bayesian decision analysis and the NUSAP approach, for which relevant references are included in the text.

The assumption deviation risk was introduced in Aven (2013d). It replaces earlier attempts to use the terms 'degree of uncertainty' and 'sensitivity' (related to the effect of the deviation on the calculated probabilities for A and C) (see e.g. Flage and Aven 2009, Eidesen and Aven 2010). The new concept is considered to be more precise in reflecting the ideas that we wish to describe. See the paper by Mosleh and Bier (1996) and Aven *et al.* (2014a) for some additional discussion on the issue of uncertainty of probabilities.

5 Risk management

Three major strategies are commonly used to manage risk: risk-informed, cautionary/precautionary and discursive strategies. In most cases the appropriate action would be a mixture of these three strategies. The risk-informed strategy refers to the treatment of risk – avoidance, reduction, transfer and retention – using risk assessments. The cautionary/precautionary strategy is also referred to as a strategy of robustness and resilience, and highlights features such as containment, constant monitoring, research to increase knowledge and the development of substitutes. The discursive strategy uses measures to build confidence and trustworthiness through the reduction of uncertainties, clarification of facts, involvement of affected people, deliberation and accountability. We refer to Renn (2008) and Aven and Renn (2010) for further details on this strategy.

In this chapter, in Section 5.1 we review common pillars for risk-informed and cost–benefit types of strategies. Then in Section 5.2 we examine a recently developed perspective on risk management, particularly highlighting surprises and black swans. In addition to the (C,U) risk perspective it builds on ideas from the quality discourse and the use of the concept of (collective) mindfulness linked to studies of High Reliability Organisations (HROs). The concept of antifragile as defined by Taleb (2012) is also incorporated into this perspective, which stresses that to obtain high performance and excellence, to a certain extent it is necessary to love variation, risk and uncertainties.

Section 5.3 addresses specific issues related to deep uncertainties regarding the events occurring and the consequences of these events, such as in preparing for climate change and managing emerging diseases: what policies and decision-making schemes should be implemented in such cases? Traditional statistical methods and tools are not suitable because relevant supporting models cannot easily be justified and relevant data are missing.

Finally, Section 5.4 includes a fundamental discussion of how to best manage the risk related to black swan events. The previous sections outlined various approaches and tools, and it is time to draw some conclusions and make recommendations.

5.1 Risk management strategies

By 'risk management' we understand all measures and activities carried out to manage risk, including the identification of threats/hazards, the assessment of risk, and risk-informed decision-making. Risk management deals with balancing the conflicts inherent in exploring opportunities on one hand, and avoiding losses, accidents, and disasters on the other. For some fixed frames it may be a goal to reduce risk (for example, the risk related to accidents in a process plant), but it has to be acknowledged that the pursuit of some benefits may increase risk over time; see Figure 5.1. If we start new activities the safety risk may increase, but this is accepted because benefits are obtained. Risk management is built on risk assessment, but it is much more than the planning, execution and use of this tool. Other activities include the establishment of goals and strategies, the establishment of roles and responsibilities, and the communication, training and development of a good culture.

The risk relates to different organisational levels. The main focus should be on the principal objectives of the organisations, which for an enterprise could be to maximise its value, avoiding health, safety and environmental (HSE) and integrity incidents.

Risk can be defined in relation to achieving goals at lower organisational levels and project executions, but it is essential that this risk management supports the work to meet the principal objectives of the organisation; see discussion in Aven and Aven (2014).

By 'risk governance' we mean risk management when there is no single authority to take a binding risk management decision; rather, the nature of the risk requires collaboration and co-ordination between a range of different stakeholders (Aven and Renn 2010). The need for suitable approaches to deal with complex risk issues has resulted in various efforts to design integrative risk governance frameworks. One such approach is the

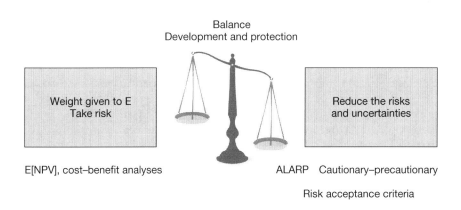

Figure 5.1 Risk management as a balancing act. E: Expected value

International Risk Governance Council (IRGC) framework which has been designed to help analyse how society could better address and respond to complex risks (Renn 2008, Aven and Renn 2010). To this end, the IRGC's framework maps out a structured approach which guides its user through the process of investigating significant risk issues and designing appropriate governance strategies. The approach combines scientific evidence with economic considerations as well as social concerns and societal values, and thus ensures that any risk-related decision draws on the broadest possible view of risk. The approach also states the case for an effective engagement of all relevant stakeholders.

To categorise the risk problems, we distinguish between three major categories; see Figure 5.2:

Small uncertainties. This category is characterised by situations and problems in which the knowledge base is strong so that accurate predictions can be made. Examples include car accidents, smoking, regularly recurring natural disasters or safety devices for high buildings. Note that the fact that the uncertainties are small does not mean that the risks are low. The possible negative consequences could be very large. The point is that it is possible to predict the occurrence of events and/or their consequences with a high degree of accuracy. As the knowledge base is so strong, black swans can for all practical reasons be ignored.

Moderate uncertainties. This category is characterised by situations between the categories of small uncertainties and large (deep) uncertainties. Some dominating explanations and beliefs exist, but the knowledge base is considerably weaker than the category of small uncertainties. Examples of situations that could be covered by this category are oil and gas exploration in Arctic areas, and some types of terrorism and sabotage.

Large (deep) uncertainties. This category refers to situations in which the knowledge base is poor, and reliable predictions cannot be made. Some hypotheses are formulated but their support is weak. Examples include many natural disasters (such as earthquakes), possible health effects of mass pollutants, and the long-term effects of introducing genetically modified species into the natural environment.

Large uncertainties may be a result of scientific uncertainties as defined by the precautionary principle. There are many ways of understanding the term 'scientific uncertainties', but most interpretations seem to be linked to a lack of understanding of how the consequences (outcomes) are influenced by underlying factors: it is difficult to establish an accurate prediction model that would lead to a precise description of a 'cause–effect relationship' (Aven 2011e).

Large uncertainties may be due to complexity, which refers to the difficulty of identifying and quantifying causal links between a multitude of potential causal agents and specific observed effects (Renn 2008). The nature of this difficulty may be traced back to interactive effects among these agents (synergism and antagonisms), long delay periods between cause and effect, inter-individual variation, intervening variables and others. Complex risk problems are often associated with major scientific dissent about dose–effect relationships or the alleged effectiveness of measures to decrease vulnerabilities. Examples of activities/systems with high complexity include sophisticated chemical facilities, the synergistic effects of potentially toxic substances, the failure of large interconnected infrastructures and critical loads to sensitive ecosystems. Black swan events may occur.

In addition to these categories, we may characterise the situation with respect to ambiguity, defined as follows:

Ambiguity (normative ambiguity) refers to different views related to the values to be protected and the priorities to be made.

For example, ambiguities can be associated with passive smoking, nuclear power, pre-natal genetic screening and genetically modified food.

For each category, a strategy can then be developed for risk assessment and risk management supported by proposals for appropriate methods and tools. See Table 5.1; its content will be discussed in coming subsections.

It is necessary to stress that the structure defined by Figure 5.2 is very simple and many risk problems do not perfectly match a single category. Nevertheless, the structure is believed to provide some useful insights about the type of problems we are facing when dealing with risk, as we will see in the coming sections.

In recent years there has been an increased focus on uncertainties in relation to risk assessment and risk management. The traditional probabilistic approach to risk assessment and management is criticised for its narrowness in the way it looks at risk and how it copes with uncertainties (see e.g. Renn 2008, Aven and Zio 2011). In response to the critique, several alternatives have been suggested. These can be grouped as follows:

(1) Replacing the probabilistic approach with other quantitative approaches, for example based on interval probabilities (typically supported by possibility theory or evidence theory) (see Appendix A and Aven *et al.* 2014a)

(2) Balancing alternative approaches, in particular the probabilistic approach and approaches that are effective in meeting hazards/threats, surprises and the unforeseen. For short, we refer to the latter approaches as 'robust approaches'. As discussed in Section 1.7, robust approaches cover cautionary measures such as designing for flexibility, and improving the performance of barriers by using redundancy, maintenance, testing etc. They also cover concepts such as resilience engineering and the concept of antifragility; see Section 5.2.3.

(3) Rejecting the probabilistic approach to risk assessment and risk management and instead relying on robust approaches.

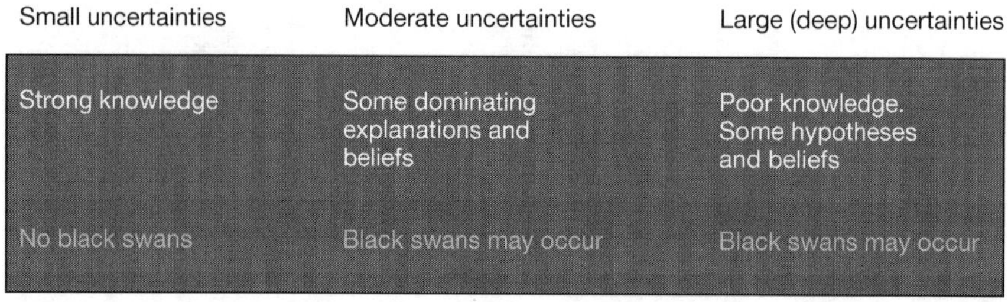

Small uncertainties	Moderate uncertainties	Large (deep) uncertainties
Strong knowledge	Some dominating explanations and beliefs	Poor knowledge. Some hypotheses and beliefs
No black swans	Black swans may occur	Black swans may occur

Figure 5.2 Different categories of risk problems (based on Aven 2013f)

Table 5.1 Risk problem categorisations and their implications for risk management (based on Aven and Renn 2010)

Risk problem category	Management strategy	Appropriate instruments
Small uncertainties	Risk informed Routine-based risk treatment (risk reduction)	Statistical analysis Risk assessments Cost–benefit analyses Trial and error Technical standards Economic incentives Education, labelling, information Voluntary agreements
Moderate uncertainties	Risk informed (risk agent)	Risk assessments, broad risk characterisations Cost–benefit analyses Tools include: • Containment • ALARP (as low as reasonably practicable) • BACT (best available control technology), etc.
	Risk informed	Risk assessments Cost–benefit analyses
	Robustness focused (risk absorbing system)	Improving buffer capacity and performance of hazard/threat risk target through for example: • High performance standards of barrier systems • Additional safety factors • Redundancy and diversity in designing safety devices • Improving coping capacity To some extent also measures mentioned for large (deep) uncertainties

Risk problem category	Management strategy	Appropriate instruments
Large (deep) uncertainties	Risk informed and caution/precaution-based (risk agent)	Risk assessments. Broad risk characterisations, highlighting uncertainties and features such as persistence, ubiquity, etc.
		Tools include: • Containment • ALARP (as low as reasonably practicable) • BACT (best available control technology), etc.
	Risk informed Robustness and Resilience focused (risk absorbing system)	Risk assessments. Broad risk characterisations. Improving capability to cope with surprises: • Diversity of means to accomplish desired benefits • Avoiding high vulnerabilities • Allowing for flexible responses • Preparedness for adaptation

In practice, alternative (2) normally applies. If we look at high risk industries such as nuclear and oil/gas, we find a mixture of probabilistic and robust approaches. It is acknowledged that probabilistic approaches have limitations in managing risk, surprises and the unforeseen, and need to be supplemented by robust approaches. For societal safety and security contexts this duality is even stronger, for example in relation to terrorism risk. Here probabilistic risk assessments are hardly used. The information provided by assigned attack probabilities is small in most cases (Brown and Cox 2011, Aven 2013c).

To confront risk and uncertainties, robust thinking is needed to a varying degree depending on the situation. This is also true when alternative quantitative approaches (approach 1) are adopted. Using interval probabilities in place of specific probability numbers can allow for more balanced judgements reflecting what is known and what is not, but the interval probability approach is not easily implemented in practice – there are many challenges – and most importantly, it cannot replace the need for robust approaches (see Section 3.3.1). No analytical quantitative approach, probabilistic or not, can make robust arrangements and measures superfluous, as such approaches will always be subject to some limitations. They can provide useful decision support, but they cannot prescribe what to do.

5.1.1 Risk informed strategies

Following the recommended risk approach, the risk assessment produces a risk description that covers identified events and consequences, assigned probabilities, uncertainty intervals and strength of knowledge judgements, as well as considerations about black swans. The assessment and its results provide insights that the decision-maker and other stakeholders can use to support decision-making and their views on relevant issues, such as choosing between alternatives, the implementation of risk-reducing measures and so on (Apostolakis 2004). The risk assessment and its description must be seen as judgements made by the analyst group and the experts that have been consulted during the assessment. The results are not objective in the sense that they exist independently of the analysts and experts. However, presuming that the assessment is conducted in a professional and scientific way it meets certain standards on quality, for example that all steps of the assessments are traceable, all assumptions used are recorded, and that all analysis principles and methods adopted are justified. The analysts and experts have high levels of competence in the field of study and in using risk assessment as a tool, and these judgements are thus of interest for the decision-maker and potentially other stakeholders – not as a prescription for what to do, but as an input to a broader process, namely managerial review and judgement. In this process, the decision-maker and potentially other stakeholders see the results of the assessment in view of its limitations, and reflect on other concerns and issues not captured by the assessment but which are still important for the decision-making. The benefits related to the activity studied, as well as strategic and political aspects, could be decisive for the decision-making but may not be captured by the assessment.

The results of the risk assessments are also used as input to broader decision analysis tools such as cost–benefit analysis, cost-effectiveness analysis and multi-attribute analysis. All these methods have in common that they are systematic approaches for organising the pros and cons of a decision alternative, but they differ with respect to the extent to which the analyst is willing to make the factors in the problem explicitly comparable. Among experts there are different views related to which of these analyses should be adopted when evaluating measures, for example safety measures. Many analysts and experts prefer multi-attribute analyses and cost-effectiveness analyses, while many economists prefer to transform all the attributes to one comparable unit using a traditional cost–benefit analysis. This raises the question of the extent to which all the attributes should be transformed to one comparable unit when evaluating measures. In the following we give a brief review of the aforementioned decision analysis tools.

TRADITIONAL COST–BENEFIT ANALYSIS

Traditional cost–benefit analysis is an approach to measure the benefits and costs of a project expressed in money. The local country's currency provides the common scale used to measure benefits and costs. Market goods are easy to transform into monetary values since the prices of market goods reflect society's willingness to pay. On the other hand, the willingness to pay for non-market goods is more difficult to estimate, and different methods such as contingent valuation and hedonic price techniques are used (Hanley and Spash 1993).

After the transformation of all attributes to monetary values, the total performance is summarised by computing the expected net present value, the E[NPV]. To measure the NPV of a project, the relevant project cash flows (the movement of money into and out of the business) are specified, and the time value of money is taken into account by discounting future cash flows by the appropriate rate of return. The formula used to calculate NPV is:

$$NPV = \sum_{t=0}^{T} \frac{X_t}{(1+r_t)^t}$$

where X_t is equal to the cash flow at year t, T is the time period considered (normally denoted in years) and r is the required rate of return, or the discount rate, at year t. The terms 'capital cost' and 'alternative cost' are also used for r. As these terms imply, r represents the investor's cost related to not employing the capital in alternative investments. When considering projects where the cash flows are known in advance, the rate of return associated with other risk-free investments, such as bank deposits, provides the basis for the discount rate to be used in the NPV calculations. When the cash flows are uncertain, which is usually the case, the cash flows are normally represented by their expected values $E[X_t]$ and the rate of return is increased on the basis of the Capital Asset Pricing Model (CAPM) in order to outweigh the possibilities for unfavourable outcomes (Copeland and Weston 1988).

A MORE PRAGMATIC VIEW ON A TRADITIONAL COST–BENEFIT ANALYSIS

A more pragmatic view of traditional cost–benefit analysis differs from a traditional cost-benefit analysis in two areas. The first is that some non-market goods can be represented by a different unit. This is done for attributes for which it is difficult to assess a monetary value, such as environmental issues or human life. However, non-market goods are not necessarily represented using a different unit. For example, we may find it appropriate to include a value for a statistical life in the analysis. The second difference is that there is no search for "correct", objective values. Searching for such values can be regarded as meaningless, as no such numbers exist; for example, the statistical life is a value that represents an attitude to risk and uncertainty, and that attitude may vary and depend on the context. Instead, the sensitivity of the conclusions of the analysis is demonstrated by presenting the results of the analysis as a function of the relevant parameters.

A result of these considerations is that a cost–benefit analysis in this category provides decision support rather than hard recommendations. The analysis must be reviewed and evaluated, since we cannot replace difficult ethical and political deliberations with a mathematical one-dimensional formula, integrating complex value judgements.

COST-EFFECTIVENESS ANALYSIS

A cost-effectiveness analysis is a decision support tool, and it has been shown to give useful support for comparisons between competing safety measures. We may think of a safety measure as cost-effective if it is (Petitti 2000):

- Less costly and at least as effective.
- More effective and more costly, with the added benefit worth the added cost.
- Less effective and less costly, with the added benefit of the alternative not worth the added cost.
- Cost-saving with an equal or better outcome.

Quantitatively, and more precisely, the cost-effectiveness can be expressed as a cost-effectiveness ratio: the ratio of change in expected costs to the change in expected effects. This type of ratio (index) usually forms the basis for the communication of cost-effectiveness between analysts and other stakeholders.

An example is used to clarify what a cost-effectiveness analysis expresses. In the example there are two competing safety measures: safety measure 1 and safety measure 2. The following notation is used:

- C_i: the investment cost associated with safety measure i (to simplify we assume that there is no annual cost associated with the safety measure).
- Z_i: the total effect related to loss of lives if safety measure i is implemented (to simplify we assume that this is the only effect of interest).
- R: the reference value. The value clarifies how much money the decision-maker is willing to pay to obtain one unit of effectiveness.

In order to compare the cost-effectiveness between the two measures, the cost-effectiveness ratio for both measures is calculated. The cost-effectiveness ratios for safety measure 1 and safety measure 2 are equal to C_1/Z_1 and C_2/Z_2 respectively. Safety measure 1 is more cost-effective than safety measure 2 if $C_1/Z_1 < C_2/Z_2$. To see whether safety measure 1 is preferred to the status quo or not, the cost-effectiveness ratio has to be compared with the reference value, R. Implementation of the safety measure is preferred to the status quo if the decision-maker is willing to pay more to obtain one unit of effectiveness than the cost-effectiveness index expresses, which means that safety measure 1 is preferred to the status quo if $R > (C_1/Z_1)$.

When the costs and effects C and Z are not known, which is the common situation in practice, the above analysis can be repeated with C and Z replaced by the corresponding expected values.

MULTI-ATTRIBUTE ANALYSIS

A multi-attribute analysis is a decision support tool analysing the consequences of the various decision alternatives separately for the various attributes. For each decision alternative, attention is given to attributes such as investment costs, operational costs, safety, environmental issues etc. For some attributes it is common to adopt quantitative analyses, while for others, such as political and social aspects, qualitative analyses are usually adopted. The total of these analyses is referred to as a multi-attribute analysis.

In a multi-attribute analysis there is no attempt to transform all the different attributes into a comparable unit. The decision-maker has to weight the different attributes. We say that the trade-offs are made implicitly.

The example with the two competing safety measures presented above is also used as a clarification of what the output of a multi-attribute analysis could be like. For each safety measure, attention is given to the investment cost (C_i) and the effect related to the loss of lives (Z_i). As the costs and effects are uncertain, attention is usually given to their expected values; $E[C_i]$ and $E[Z_i]$. The appropriateness of each safety measure is considered by the decision-maker through the values $E[C_i]$ and $E[Z_i]$. In the overall evaluation of the 'goodness' of each safety measure, attention will also usually be given to the uncertainties, as the expected values could produce poor predictions of the real outcome. Risk considerations beyond expected values are required.

EXPECTED UTILITY THEORY

This theory, which is briefly discussed in Section 2.2.1, is the ruling theoretical paradigm for decision-making under uncertainty, which states that the decision alternative with highest expected utility is the best alternative. In mathematical terms, the expected utility is written as $Eu(X)$, where u is the utility function and X is the outcome expressing a vector of different attributes, for example costs and the number of fatalities. Through expected utility theory we may reflect that we dislike negative consequences so much that these are given more weight than is justified by reference to the expected value. The decision-maker's attitude towards risk is referred to as 'risk averse', which is the standard behavioural assumption. The decision-maker can also be a 'risk seeker' or 'risk neutral'. Mathematically, these terms are defined as follows: we call the decision-maker's behaviour risk averse if $Eu(X) < u(EX)$. The behaviour is risk neutral if $Eu(X) = u(EX)$, and they are risk seeking if $Eu(X) > u(EX)$. We will not repeat the rationale for the expected utility principle, but it has validity under very reasonable considerations for logical and consistent behaviour. See for example Savage (1962), von Neumann and

Morgenstern (1944), Lindley (1985) and Bedford and Cooke (2001). However, the practical use of the theory is limited, since the theory requires that the decision-maker specifies the utility function, which is not realistic in typical risk problems. The specification of the utility function is also very difficult; see Section 2.2.1 and Aven (2010c).

5.1.2 Cautionary/precautionary strategies

The cautionary principle expresses that, in the face of uncertainty, *caution* should be a ruling principle, for example by not starting an activity, or by implementing measures to reduce risks and uncertainties, even if these measures are inefficient when seen from a purely economic viewpoint. Of course, the level of caution adopted has to be balanced against other concerns such as costs, but to be cautious goes beyond balancing the expected benefit of risk reductions expressed in monetary terms against expected costs. Many emergency preparedness measures in our society, implemented to protect people and the environment, can be considered justified by reference to the cautionary principle.

The precautionary principle may be considered a special case of the cautionary principle, as it is a legal basis for cautionary actions in the face of scientific uncertainties, as was briefly discussed in relation to deep uncertainties in Figure 5.2.

Using uncertainty characterisations and risk scenarios, analysts are able to identify events that could be covered by a cautionary approach even if the assigned probability of such an event is low. At the same time, precaution does not mandate the rejection of risk whenever the background knowledge is poor or the extent of negative consequences is uncertain. It is certainly not prudent to protect oneself against all possible events. The precautionary principle advises risk managers and regulators to introduce risk management measures (and, in extreme cases, bans) only for those situations in which there is no conclusive evidence about the cause–effect relationship. A primary thrust of precaution is to avoid irreversibility.

Hence, the critical criteria for judging whether the precautionary principle should apply are both the possibility of severe impacts and the existence of scientific uncertainties. Consider the issue of having continuous petroleum activity in the Barents Sea (Aven and Renn 2012). Clearly the first condition is met – the implications could be serious. The second condition, concerning whether there are scientific uncertainties related to this activity, needs more exploration.

Consider the consequences of an oil spill on fish species, and let Z denote the recovery time for the population of concern, with Z being infinite if the population does not recover. We may conclude that there is scientific certainty according to our understanding of this concept if we can establish a function (model) G such that Z equals G(X) with high confidence, where $X = (X_1, X_2, ...)$ are underlying factors influencing Z. Such factors could relate to the possible occurrence of a blowout, the amount and distribution of the oil spilled on the sea surface, the mechanisms of dispersion and degradation of oil components, and the exposure and effect on the fish species. For values of X, we can use G to predict the consequences Z.

Scientific consensus in this sense does not mean that the consequences (Z) can be predicted with accuracy when the analysis is not conditioned on the Xs. Unconditionally, the consequences (Z) are, of course, uncertain, and this uncertainty is defined by the uncertainties of each component of X. In this case there are considerable uncertainties about some components of X, and the model G is disputed among experts. For example, many biologists may conclude that there is a lack of fundamental understanding of the underlying phenomena concerning the effect on the fish species, and therefore an accurate prediction model G cannot be established. However, others argue that such a model can be constructed in a valid format; potential surprises can for all practical reasons be ignored. The two frames co-exist in parallel and there is no argument to refute the one in favour of the other. Looking at the certainties on one side and highlighting the uncertainties on the other are two perspectives that relate to the same object like the two sides of a coin. On a meta-level, one could conclude that precaution is warranted if at least one trustworthy camp claims that there are scientific uncertainties that need to be taken into account. What this means in practice is difficult to delineate and requires prudent judgement. For example, the highest institution in the Norwegian Church referred to the precautionary principle when they requested the Norwegian government not to permit year-round petroleum operations in the Barents Sea some years ago.

For a large proportion of risk-informed decisions there will be some degree of scientific uncertainty. Hence, the question arises: at what point should precautionary measures be taken? At what point is the lack of scientific certainty a trigger for initiating risk management measures? How accurate does the model G need to be to justify such measures?

There are no clear answers to these questions. Different people and parties would judge these issues differently. There are no sharp limits stating that a specific level is not acceptable and that the precautionary principle should apply. Hence, referring to the precautionary principle implies a judgement expressing that we find the lack of scientific uncertainty to be so significant that precautionary measures are required. In extreme cases, the whole activity should be terminated. In other cases, additional risk-reduction measures are sufficient. An extended risk assessment producing predictions and uncertainty characterisations can provide input to inform such a judgement.

In addition, stakeholders' and people's risk perceptions may influence the judgement of the decision-makers and their assignment of the importance that they attribute to various aspects of the risks and benefits. The relative weights depend not only on the degree of value violation or fulfilment that corresponds with the respective dimension, such as job creation, marine pollution or occupational health, but also on the degree of uncertainty that such a dimension will be violated or met. The more uncertain the consequences on each of these dimensions, the more decision-makers are motivated to apply the precautionary principle, even if the expected value for each risk dimension is quite low.

Many parties evoked the precautionary principle when requesting the banning of continuous petroleum activities in the Barents Sea. However, their judgements were probably more justified by reference to the more general cautionary principle than the precautionary principle. Their concern was not so much about scientific uncertainties as the fact that uncertainties exist: an oil spill could occur causing severe environmental damage. We could calculate fairly small probabilities for such events and scenarios, but they are still possible.

The Norwegian government gave strong weight to the cautionary or precautionary principle in their decision-making, since not all fields were opened for year-around operations. The particular decisive factor is not really important, as long as the arguments are made clear: there are risks and uncertainties associated with the activity and these were considered so significant for some areas in relation to the values at stake that year-round activities were unacceptable at this stage.

To some parties and stakeholders the activities were judged on the basis of the perceived uncertainty and the applicability of the cautionary or precautionary principle. However, uncertainty is only one reason for the dissenting view on acceptability. A second reason is the assessment of ambiguity attached to weighing different attributes (risk, income, etc.) in the risk–benefit balancing act. We will discuss this aspect in more detail in the next section.

RISK ACCEPTANCE AND TOLERABILITY CRITERIA

Risk acceptance and tolerability criteria are often used to control the risk in relation to some attributes, such as safety for human lives. For example, this was the case in the LNG study referred to in Section 4.2, where the operator defined risk acceptance limits based on f-n curves and maximum individual risk numbers. If the risk assessment shows probability figures below the defined limits, the risk is considered acceptable, whereas if the probability figures exceed the limits, the risk is considered unacceptable. Such criteria must be used with care because they can easily lead to the wrong focus, meeting the criteria instead of finding the overall best arrangements and measures (Aven and Vinnem 2005). For the practical execution of risk management activities, it is not difficult to see that some types of criteria may be useful in simplifying the decision-making process. In the design of the LNG plant, it was attractive to have some concrete reference values that could be used to make judgements about the acceptability of the risk. However, a simple reference to the probability assignments would obviously not be justifiable, since as we have seen, risk is much more than probability numbers.

To adjust the approach, and reflect aspects of risk other than probability, the following procedure has been suggested (Aven 2013d; see also Table 5.2):

(1) If risk is found to be acceptable according to probability with large margins, the risk is judged to be acceptable unless the strength of knowledge is weak (in this case the probability-based approach should not be given much weight).
(2) If risk is found to be acceptable according to probability, and the strength of knowledge is strong, the risk is judged to be acceptable.
(3) If risk is found to be acceptable according to probability with moderate or small margins, and the strength of knowledge is not strong, the risk is judged to be unacceptable and measures are required to reduce risk.
(4) If risk is found to be unacceptable according to probability, the risk is judged to be unacceptable and measures are required to reduce risk.

The approach relies on cautionary thinking.

ALARP

The above scheme balances the need for a simple practical procedure with the necessity to reflect the strength of the background knowledge in the judgements. In addition, risk-reducing processes are commonly required, as in the LNG case, and this is often considered in line with the ALARP principle (ALARP: As Low As Reasonably Practicable). This principle is based on the idea of gross disproportion and states that a risk-reducing measure shall be implemented unless it can be demonstrated that the costs are in gross disproportion to the benefits gained. Inspired by the ideas of Aven and Vinnem (2007), this can be implemented as follows when considering the need for a specific risk-reducing measure:

Table 5.2 Adjusted procedure for use of risk acceptance criteria in view of considerations of the strength of knowledge (based on Aven 2014c)

Probability-based justification	Above limits	Unacceptable risk	Unacceptable risk	Unacceptable risk
	Small margin below	Unacceptable risk	Unacceptable risk	Acceptable risk
	Large margins	Further considerations needed	Acceptable risk	Acceptable risk
		Poor	Medium	Strong
	Strength of knowledge			

(a) Implement the measure if the cost is small.
(b) Implement the measure if formal cost–benefit analyses or cost-effectiveness analyses show that the measure is justifiable.
(c) Also, if a and b are not justified, consider implementing the measure if the strength of knowledge is poor or medium, or the measure can reduce black swan risks or have other positive effects on robustness and resilience (cf. Renn 2008, Hollnagel *et al.* 2006, Aven 2014c).

DISCUSSION

In many cases there could be conflicts of interest, as in the LNG case. The operator would like to prove that the plant is safe, and using probability-based criteria would be an efficient approach to do this. The safe status is ensured if the probability numbers produced by the risk assessment are sufficiently small. The problematic issue of uncertainties has been basically eliminated. As commented by Tickner and Kriebel (2006), there is a tendency for actors not to talk about uncertainties underlying the risk numbers: acknowledging uncertainty can weaken the authority of these actors. However, as argued in this book, that risk is more than assigned probabilities, and the simple procedure of using probability-based risk criteria cannot be justified. Lack of knowledge, and surprises, are aspects of risk and they need to be reflected in the way risk is assessed and treated.

An approach such as the one recommended here is thus expected to be met with some resistance. From a simple business perspective, it may not serve the interests of the company to add these aspects of uncertainty into the risk description. However, for third parties (for example, people living close to the LNG plant who are fighting against the development of the plant) the focus on the uncertainties could be seen as beneficial to their case. Hence, it is essential that the agency is clear on its role and can steer the risk management and overall risk management regime in an adequate way. The risk assessments need to be framed and conducted in a way that provides a fair and professional study of the risks and uncertainties.

Here is another potential conflict, since the probability-based approach is also attractive to the authorities (safety and security agencies) because compliance can more easily be followed up compared to a system which includes more or less structured judgement processes. The agency may acknowledge the need for an approach that reflects the knowledge and surprise aspects, but it is 'forced to' accept the probability-based approach because it is more easily supervised and controlled. The fact that modern safety management is based on internal control, meaning that the company has full responsibility for its activities, makes it very difficult to impose procedures that are not easily implemented and can be followed up in practice. A good example is the use of the ALARP principle in the Norwegian oil and gas industry. The principle is a part of the regulation regime, but the industry struggles to use it effectively, as it needs to be based on overall judgements and not mechanised procedures (Khorsandi *et al.* 2012).

Clearly, risk reduction and a high safety level in such a setting can only be achieved if the company has an ambition in this direction. The knowledge aspects and the black swan list can to a large extent be ignored or given little attention if the focus is cost reduction. The agency could force the operators to conduct risk assessments that cover these aspects, but it would not lead to different decisions unless the company believes in the importance of taking these issues into account.

The above analysis very much fits the LNG case. For the national risk assessment case addressed in Section 4.2.1, the discussion is somewhat different. The assessment is not directly used for supporting decisions and prioritising measures; rather, it is meant to provide some general insights about the risks we are facing and how 'large they are', as a general background for societal safety and security planning. Having said that, it is obvious that adding the knowledge and surprise aspects of risk, as here recommended, could dramatically change the information provided by the risk assessment and in turn affect the decision-making. Some risk events may have quite a low risk, as seen from the probability and expected consequences perspective, but considerable risk when adding the uncertainties in the consequences, the strength of knowledge and the black swans' judgements. Making a judgement as to whether the risk of an event is high or not clearly cannot be done by a simple formula, because the dimensions are many, with various attributes. As such, the information provided by the results of the risk assessment merely forms one of many inputs to the overall judgements that bureaucrats and politicians need to make in societal safety and security planning. The assessment risk-informs the decision-makers and other stakeholders, but it does not show the right way of looking at risk and which decisions are needed. The decisions are for the right actors to make, reflecting all relevant issues and adequately balancing different concerns, as highlighted already many times.

WHY RISK ACCEPTANCE CRITERIA NEED TO BE DEFINED BY THE AUTHORITIES AND NOT THE INDUSTRY

As a point of departure for the reflections here, we refer to the regulations of the petroleum activities on the Norwegian Continental shelf. The Health, Environmental and Safety (HES) regulations in the Norwegian petroleum sector is founded on internal control (Hopkins and Hale 2002). This means that the licensees have full responsibility for ensuring that the petroleum activities are carried out in compliance with the conditions laid down in the legislation, and the authorities' supervisory activities aim to ensure that the licensees' management systems cater adequately for the safety and working environment aspects in their activities.

The HES regulations state that the operator has a duty to formulate their own risk acceptance criteria (the upper limits of acceptable risk), which are in line with the internal control principle. This practice of formulating risk acceptance criteria is in contrast to what is done in many countries and industries – for example in the UK, where the risk acceptance criteria (tolerability limits) are formulated by the authorities. This issue has been discussed by Abrahamsen and Aven (2012). They question to what degree the Norwegian approach can be justified from a societal safety point of view.

There are many aspects to consider – including fundamental economic principles that govern the operators' willingness to invest in safety. Of major importance here is the fact that an operator's activity will usually cause negative externalities to society. An externality is an economically significant effect due to the activities of an agent/firm that does not influence the agent's/firm's production, but which influences other agents' decisions (Varian 1999). An accidental event may, for example, lead to loss of lives, environmental damage etc., which are not fully taken into consideration by the firm when managing its activity.

The discussion by Abrahamsen and Aven (2012) is founded on the expected utility theory, as was briefly reviewed in Section 5.1.1. Using this theory the authors show that risk acceptance criteria formulated by the operators would not in general serve the interest of society as a whole. The main reason is that an operator's activity will usually cause negative externalities to society. The increased losses for society imply that society wants to adopt stricter risk acceptance criteria than those an operator finds optimal in its private optimisation calculations. The conclusion is that if risk acceptance criteria are to be introduced as a risk management tool, they should be formulated by the authorities and not by the operators.

Also, today the HSE regulations issued by the Petroleum Safety Authority Norway (PSA) refer to a type of concrete risk acceptance criteria – the so-called 1×10^{-4} criteria for safety functions (a maximum of a probability of 1×10^{-4} for specific types of accidental loads) to be applied in the early design of petroleum installations (Aven and Vinnem 2005). These criteria are introduced by the PSA to ensure a minimum safety level, and they demonstrate that whenever it is found appropriate, the authorities impose specific requirements on industry. There is obviously a balance to be struck by the authorities concerning when it is indeed necessary to define specific requirements (as this may be considered to be in conflict with the basic principle of internal control), but according to Abrahamsen and Aven (2012), the present situation calls for action. The overall ambitious safety goals set for the industry by the Norwegian government (Aven *et al.* 2010) are threatened by the existing regime.

The use of risk acceptance criteria (risk tolerability limits) is intuitively appealing. First come the criteria – typically expressed by probabilities – followed by the analysis to see whether these criteria are met. Finally, according to the assessment results the need for risk-reducing measures is determined. However, as discussed in Aven and Vinnem (2005, 2007), a closer look at this practice reveals several problems. First, we have the general problems related to management by objectives, as will be discussed in Section 5.2. The point is that the introduction of pre-determined criteria may give the wrong focus – meeting these criteria rather than obtaining overall good and cost-effective solutions and measures. Practice shows that when such criteria are defined, the focus is on meeting them and improvement processes are not stimulated. Second, the risk is commonly reduced to probability by this approach, and as we have discussed extensively in this book, such a perspective on risk is too narrow. The strength of knowledge supporting the assigned probabilities always has to be added to the risk description. Hence the practice of acceptance and non-acceptance according to simple judgements of probability exceeding or not exceeding particular threshold values cannot be justified.

As discussed above, it is possible to adjust the schemes for using risk acceptance criteria by also reflecting the strength of knowledge, but it is also possible, as shown by Aven and Vinnem (2005, 2007), to manage risk without such criteria.

In practice there will always be a need for some guidelines to ensure consistency and transparency in the decision-making processes, but these should not be mechanistic using probabilistic accept/reject limits. Such simple risk rules cannot be justified and are in conflict with the second principle of the collective mindfulness concept 'reluctant to simplify' (see Section 1.2 and 5.2.1). Instead the management regime can be built on the ALARP principle as outlined above, with suitable detailed requirements set by the authorities on critical issues – as we see today in the oil and gas industry, for example. When considering risk, the knowledge and uncertainty aspects always need to be reflected, and ignoring these dimensions introduces an arbitrariness in the management processes that is unacceptable. It has to be acknowledged that handling risk is not a mechanistic process. Practical guidelines for the treatment of risk and related decision-making need to reflect this.

5.1.3 Risk communication

Risk communication can be seriously hampered if the risk assessment and management lack a proper platform. On the other hand, if a solid platform is in place, it is much more likely that risk communication will work effectively as the premises for the dialogue are clear. As argued in Veland and Aven (2013), the main barrier to good risk communication is not the layman's poor understanding of the risks and the risk assessment tools, but the risk analysts who have not done their job in a professional way.

In practice we often see that the analysts seek to meet the challenge of decision-makers who lack risk analysis competence by trying to keep analyses simple and avoiding discussions of uncertainties (Aven 2011a, p. 125). However, risk may thus be poorly described, since uncertainty is an important dimension of risk. Even if the decision-maker lacks fundamental training in risk, the risk communication can be informative, provided that the analyst does his/her job in a professional way. Managers and politicians are able to relate to and deal with uncertainties and risk; these tasks are largely what their job is all about – making decisions under uncertainty and risks. Managers are usually well-equipped people who will quickly understand what is at stake and what the key issues are if the professionals can do their job. The problem is rather that the analysts are not able to report the uncertainties and present them in an adequate way.

This book argues for the (C,U) risk perspective. However, independent of the perspective adopted, the requirement of professionalism in relation to a scientific platform is fundamental. If a concept is introduced it must be given a meaningful definition and interpretation. That is unfortunately not the case today in many situations (Aven 2011b). Even when an 'objective perspective' of risk is adopted, meaningful communication can take place if due consideration is given to the understanding of the concepts introduced and the uncertainties involved. There has been a tendency for risk analysts and decision-makers coming from the 'objective perspective' to conceal uncertainties, and we see that this is still often the case (Aven 2011b). On the other hand, in following (C,U) arguments, we may experience the other extreme, that too much focus is placed on the uncertainties. It is for the risk assessment discipline to decide what the proper level is, through the establishment of proper scientific principles and methods. More research is required on this issue, but equally important is the recognition among risk professionals that meaningful risk communication relies on a solid basis. Improvements have to be made in this area to equip risk analysts and also decision-makers with the necessary competence and understanding for these matters.

5.2 An extended risk and performance perspective

The risk-based approaches incorporate risk assessments, but they need to be extended and have a broader scope than the standard probabilistic analysis commonly seen in text books and practice today, as indicated in the previous section. A focus on knowledge building, transfer of experience and learning represents an important means to manage the risk related to surprises and black swans, by obtaining an improved understanding of relevant systems and activities, models and the ability to predict what is coming. To provide a suitable foundation for such improvements, we need a platform that incorporates adequate concepts, assessment and management principles and methods. This is a research issue, and some ideas for how such a platform can be defined are presented by Aven and Krohn (2014). The idea is to integrate the conceptual framework of the (C,U) risk perspective, as outlined in previous sections, with associated assessment and management principles and methods, and add theories and practical insights from other fields specifically addressing the knowledge dimension and black swans. In Aven and Krohn (2014), two areas are highlighted; the first of these is the collective mindfulness concept linked to High Reliability Organisations (HROs), with its five principles: preoccupation with failure, reluctance to simplify, sensitivity to operations, commitment to resilience and deference to expertise. There is a vast amount of literature (see e.g. Weick and Sutcliffe 2007, Weick *et al.* 1999, Le Coze 2013, Hopkins 2014) providing arguments for organisations to coordinate their efforts in line with these principles to obtain high performance (high reliability) and effectively manage risks, the unforeseen and potential surprises. The second area relates to the quality discourse with its focus on variation, system thinking and continuous improvements. In addition we have added the concept of antifragility (Taleb 2012).

5.2.1 Collective mindfulness

For the mindfulness concept with its five characteristics, the talk example of Section 1.2 points to many important issues, for example the focus on signals and early warnings. A good example of the importance of being sensitive to operations and adequately reading signals and warnings is the *Deepwater Horizon* accident, in which a worker overlooked a warning about the blast – he did not alert others on the rig as pressure increased in the drilling pipe, a sign of a possible 'kick' (*Financial Post* 2013). A kick is an entry of gas or fluid into the wellbore, which can set off a blowout.

Looking at the major accidents we have had in the oil and gas industry, a typical characteristic is that there have been strong indications of something being wrong, but due to a poor understanding of risk, the necessary actions were not taken. It is a challenge for risk management to take into account all relevant warnings and signals, and conclude what are 'false alarms' and what are not; refer to Section 4.3. To make such judgements, we need to rely on some type of risk and uncertainty assessments, but their form and basis are not straightforward. There is a need for further research on this issue; central here is the understanding and description of how risk develops over time, as well as the authority and weight to be given to the expert judgements. Our way of thinking about risk may provide valuable input to the judgements to be made about critical situations and events, in the form of more informative characterisations of risk and uncertainties than the standard perspectives provide. However, these characterisations cannot remove the need for value judgements by relevant persons, related to how to give weight to the different types of uncertainties and weigh them against other aspects, including profit issues.

The 'reluctant to simplify' principle of mindfulness means in this case that we will not allow the judgement of risk to be based only on simple risk matrices as used in job safety analysis, which is a common risk assessment tool in the oil and gas industry, for example (Leistad and Bradley 2009). Risk is more than probabilities and expected consequences as discussed above. As clearly demonstrated by the study by Leistad and Bradley (2009), the current job safety analysis practice has severe weaknesses in its ability to reveal risk contributors and create a proper understanding of risk.

The fourth collective mindfulness principle, commitment to resilience, should make John in the talk example focus in advance on training and measures that prepare him for special situations and also surprises. However, in practice perfection can seldom be achieved and events may occur that are not easily dealt with. The resilience preparation can still be useful in guiding adaptive risk management when such a surprise appears.

Experts are needed to make judgements about risk and uncertainties, and we remember the fifth characteristic of the mindfulness concept, deference to expertise, which stresses the importance of allowing people with the competence to make the important judgements and decisions in critical operational situations. It may be preferable to let an experienced and competent operational manager make a decision in the case of an emergency on an offshore installation than wait for the platform manager, who may lack practical experience. Seemingly, this type of reasoning is in conflict with the standard thinking in risk management, that various assessments are conducted to support adequate decision-making at the proper authority level. However, decisions are of different types; those at the sharp end (close to operation) could require immediate action and are completely different from those at the blunt end, which allow for considerable deliberations before a decision is made. Nonetheless, whoever makes the decision, there is always a need to see beyond the assessments available, to reflect on their scope and limitations and take into account other aspects and concerns, to the extent that time allows. It is important to recognise the considerations, the managerial review and judgement, and their content should always be questioned to ensure some level of traceability and structure for the decision-making.

5.2.2 Quality discourse

In addition, the use of ideas from the quality discourse, with its link to the concepts of 'common-cause variation' and 'special-cause variation' (see Section 3.1.1), and the continuous focus on learning and improvements, are added.

We have already commented on the continuous improvement related to the trial-talks of the example in Section 1.2. However, there are many other features and we will address some of them here. First, we have the thesis that most management activities cannot be measured (Deming 2000) in some objective or inter-subjective way. For example, the benefit of training cannot be measured – the cost, yes, but not the benefits. It is a myth, Deming says, a costly myth, that "if you can't measure it, you can't manage it". For our talk example, John may assign probabilities, but they will be subjective and strongly dependent on the assumptions on which the assignments are based. There will be considerable uncertainties related to a number of issues (for example, the atmosphere in the room and the type of questions), and hence it is essential to be trained and prepared in such a way that both normal variation and surprises can be dealt with effectively. Emphasising the cautionary principle, robustness and resilience is a cornerstone of the argumentation in this regard.

Second, we have the quality field's concern related to using management by objectives (MBOs). This approach is well established in industry and the public sector. The idea is to formulate objectives and then assess the performance of the activities in relation to these objectives. In this way, risk can be defined in relation to the deviation between the objectives and the actual performance. It is also a common practice to parcel out the overall organisational objectives to the various components or divisions. The usual assumption is that if every component or division accomplishes its share, the whole organisation will accomplish the overall objectives (Deming 2000, p. 30). Of course, the problem with this approach is that there are interdependences, and the efforts of the various components do not add up. Meeting one goal may lead to less flexibility with respect to other dimensions, and the overall gain is lost. As for the second principle of the mindfulness concept, reluctance to simplify, we need to have a focus on the overall performance and risk of the activity, to cover the total picture. For John's talk example, objectives can be formulated for different aspects or phases of the talk, for example the opening, the closure, the use of humour, the use of voice etc. (see Section 4.1.1 and Table 4.1), but clearly care has to be shown when focusing on each objective in isolation, since a higher performance level in one attribute could be negative for another. Top scores for humour may result in a bad talk overall, because the audience may find the speaker focuses too much on entertainment and too little on the talk's scientific content.

Third, the quality field emphasises the need for work on methods aimed at improving processes rather than focusing on setting numerical goals. The point here is that a goal alone accomplishes nothing. It easily leads to distortion and faking (Deming 2000, p. 31). What becomes important is meeting the goal, not the long-term losses that it could cause. For example, let us think of a situation where a goal is formulated as a specific probability (say 95%) for having a successful talk, as assigned by some of John's colleagues. Clearly, such a number would contribute little to the success of John's talk, without going into the method of how to obtain the numbers. The quality field answer is to focus on understanding and improving the processes that lead to failure, deviations etc. This leads us to the fourth issue raised by the quality movement.

This concerns the distinction between common causes of variation and special causes of variation, as discussed in Section 3.1.1. The former variation relates to stable processes, where accurate predictions can be made, whereas the latter covers unstable and unpredictable performance. A key challenge is to discern when we have a stable process and when we do not. In our example, if John is an experienced speaker, having given a huge number of talks, he knows that there will be variation in the audience, in his state on that particular day and so on. This reflects common-cause variation. His experience has prepared him for this type of variation, but he also needs to be prepared for special-cause variation, which can be seen as unforeseen events and surprises compared to his established routines. One day there might be a person in the audience who is hostile towards something he is saying or for any other reason. Probably John would not be prepared for such an event. Given our new way of thinking about risk, he could also experience special-cause variation: the concealed uncertainties in assumptions, the unforeseen events and surprises.

Fifth and finally, it is the thesis of the quality field that knowledge is built on theory as discussed in Section 4.3.

5.2.3 Antifragility

Nassim Taleb's antifragility concept has received considerable interest in the media and on the internet recently. For Taleb, the antifragility concept is a blueprint for living in a black swan world, the key being to love variation and uncertainty to some degree, and thus also to embrace errors. Here it is argued that Taleb's antifragility concept can be seen as an ideal state, where we are exposed to some level of variation and uncertainties but we are protected from adverse events. However, such a state is unreachable, and we need guidance and ways of supporting decision-making, i.e. risk management, but we do not need purely probability-based methods (which are found to be useless by Taleb). Properly designed and run, risk management has a role to play, but only if it is acting in line with fundamental principles such as robustness, resilience and antifragility, making us able to withstand stressors and become better and better. Therefore, the concept of antifragility fits nicely into our way of understanding risk and obtaining excellence in performance.

In his book, Taleb (2012, pp. 4–5) writes:

> ...by grasping the mechanism of antifragility we can build a systematic and broad guide to nonpredictive decision making under uncertainty

in business, politics, medicine, and life in general – anywhere the unknown preponderates, any situations in which there is randomness, unpredictability, opacity, or incomplete understanding of things.

It is far easier to figure out if something is fragile than to predict the occurrence of an event that may harm. Fragility can be measured; risk is not measurable (outside the casinos or the minds of people who call themselves "risk experts"). This provides a solution to what I've called the Black Swan problem – the impossibility of calculating the risk of consequential rare events and predicting their occurrence. Sensitivity to harm from volatility is tractable, more so than forecasting the event that would cause the harm. So we propose to stand our current approaches to prediction, prognostication, and risk management on their heads.

In every domain or area of application, we propose rules for moving from the fragile toward the antifragile, through reduction of fragility or harnessing antifragility. And we can almost always detect antifragility, (and fragility) using a simple test of asymmetry: anything that has more upside than downside from random events (or certain shocks) is antifragile, the reverse is fragile.

According to Taleb, prediction and risk management should be replaced by processes moving us from states of fragility to antifragility, but is there in fact a conflict? Do we not need prediction and risk management in this process to move us in the right direction? Below we argue for the confirmation of the latter question, and the thesis that the concept of antifragility represents a useful contribution to the practice of risk management. However, we need to see beyond the perspectives on risk management addressed by Taleb (2012) to obtain this. The main goal of risk management is not to accurately estimate rare event probabilities but to reveal and assess uncertainties, and make adequate decisions under uncertainty. Before we contrast risk management and the concept of antifragility, and the related concepts of fragility and robustness, a brief review of this concept is provided.

THE ANTIFRAGILITY CONCEPT AND THE RELATED CONCEPTS OF FRAGILITY AND ROBUSTNESS

To explain the antifragility concept as defined by Taleb, Figures 5.3 to 5.5 are illustrative. Figure 5.3 shows a robust/resilient system. The system is subject to shocks and stressors, but the consequences are relatively small. The system is characterised by a rather narrow frequency distribution. We can say that the system is under control, the uncertainties are small. Think of a production system in which the failure of a unit is fixed quickly in order to resume production.

Figure 5.4 shows a fragile system. This system is not under control. Here we may experience large negative consequences; the frequency distribution of events has considerable mass on negative values. In the example of the production system, the result of a failure could be a complete shutdown of the process lasting several months.

Figure 5.5 shows an antifragile system. The extreme consequences are now only positive – the frequency distribution places heavy weight on largepositive values. The antifragile system is rewarded with good results and protected from adverse events. In the production system example, failures are fixed, but there is also an improvement process leading to better performance.

In the following section we will look more closely at these ideas and relate them to fundamental concepts and principles of risk management.

RISK MANAGEMENT AND ANTIFRAGILITY

The above description should make it clear that antifragility is an ideal state, as there are no severe negative consequences, only positive ones. Look at Figure 5.5. Here we see small perturbations around the reference value

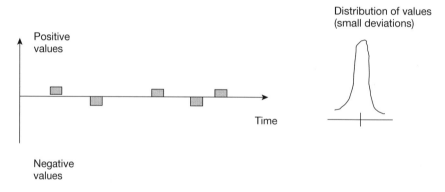

Figure 5.3 Illustration of the robust/resilient system (based on Taleb 2012)

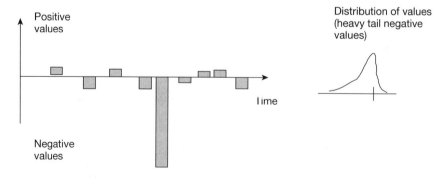

Figure 5.4 Illustration of the fragile system (based on Taleb 2012)

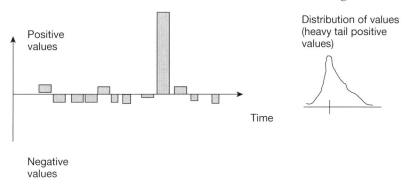

Figure 5.5 Illustration of the antifragile system (based on Taleb 2012)

(typically zero) and one extreme positive outcome. This is a realisation of an antifragile system. It is hard to think of any system in real life that is like this, which guarantees no extreme negative results. However, we can think of this state as one to aim for. A real system is antifragile to some extent, and we can think about processes to improve the system in the direction of being more antifragile. This leads us to the question of how we can measure antifragility.

Taleb argues that we can estimate and measure fragility and antifragility because they are part of the current property of an object (Taleb 2012, p. 9). He gives a number of examples related to fragility – for example, it is possible to state with high confidence that one structure is more fragile than another should a certain type of event occur, and you can easily tell that your grandmother is more fragile to abrupt changes in temperature than you.

FRAGILITY, VULNERABILITY AND RESILIENCE

Fragility is commonly understood as the quality of being 'easily broken', 'damaged', or 'destroyed', and from the two examples referred to above we get a feel for what fragile means and how it can be described or measured. It is more difficult for more complex systems, for example an offshore petroleum installation. However, for such systems, we can also obtain a lot of measurements that provide information about their fragility, for example accidents showing how relatively minor events (such as a small gas leakage) have resulted in total installation losses.

In the professional risk management context it is more common to address the vulnerability and resilience of a system than its fragility. From a practical point of view, it is hard to see that the fragility concept offers anything that is not also captured by the vulnerability and resilience concepts. Vulnerability and resilience also relate to consequences of stresses. If a system is easily broken, it is vulnerable and not resilient. As we will show, the opposite also holds under some conditions. If the consequences of some stresses are (likely to be) severe, i.e. the system is vulnerable, the system is also easily damaged by this stress; it is fragile with respect to this type of stress. For example, a person may be vulnerable in relation to people talking about his hair, which means that his mind is fragile in relation to this type of talk. However, the concept of fragility is not in general associated with specific types of stress, and this leads us to the resilience concept. If the system is *not* resilient it is *not* able to sustain its function in the face of any type of stress. The body can be non-resilient towards different types of viruses, but does this mean that the body is easily broken, fragile? The answer depends on which types of stresses we include. If our focus is on these viruses, this non-resilient state implies fragility, but not if these viruses are extremely uncommon and we take an overall body health perspective. We are again back to the issue of which types of stress to include in the considerations. For all these three concepts we have to be specific with respect to the set of stresses (known or unknown). If the set is the same, we can simply refer to one concept, for example vulnerability (in line with the ideas outlined in Aven [2011c]).

To formalise these concepts, see the general set-up of Section 2.3. Let A be the stress (coming from a set of stress types S which could be known, unknown or both) and C the associated consequences (for example, reflecting the magnitude of the damage). Then we define vulnerability (in this wide sense) as (C,U|A,S), i.e. as the combination of the consequences C of A and associated uncertainties U (what will C be?), given A and S.

To describe or measure the vulnerability we select suitable characterisations C' of C (for example, the number of fatalities), and a measure (in a wide sense) Q of the uncertainties. In general terms we can then describe vulnerability as:

$$(C',Q,K|A,S)$$

where K is the knowledge on which C' and Q are based. The most common measure Q is probability, but others also exist, including imprecise probabilities and qualitative approaches.

An assessor will judge the vulnerability given A and S to be high if he/she finds (C',Q,K|A,S) to be high.

We may define fragility in this framework by a suitable specification of the various elements. In the simplest case of a structure subject to stress, the degree of fragility can be linked to the time of failure (more precisely the expected failure time, or more generally the probability distribution of this time) given specific loads. If the times to failure are typically low, the structure is judged to be fragile. We see the similarity with vulnerability judgements. The fragility measure is also a measure of vulnerability.

Robustness (and resilience) are commonly considered to be antonyms of vulnerability (e.g. Starossek and Haberland 2010, Scholz *et al.* 2012, Aven 2011c). However, according to Taleb and Figure 5.3 above, robustness is an ideal state in the sense that we can disregard the extreme consequences. In line with the perspective adopted in Section 2.3, we have high robustness if the vulnerability description (C',Q,K|A,S) is judged to be low. However, this does not exclude the fact that extreme consequences may occur – in a complex system we cannot exclude scenarios not thought of by the analysis group, or those which were given little attention because of low assigned probabilities, even when the system is considered robust. Figure 5.3 is problematic because it ignores these types of events. The distribution of values seems to represent a true variation also applicable for the future, but this is not realistic in complex situations. The variation may be based on a substantial amount of historical data, which may be highly relevant for the future and capture most contributors to variation. However, it may also be based mainly on expert judgements, and it may be strongly influenced by those who perform the analysis. In any case, we need to consider the analysis as a subjective (or inter-subjective) measure of uncertainty and variation. Surprises may occur relative to these data, this analysis and these judgements.

Taleb warns us against the experts' risk estimates – they cannot be accurately derived, and black swans are not reflected. However, surprising events (which may be also referred to as black swan events) may also occur in relation to the judgements made about robustness (and vulnerability and fragility). The point is that uncertainties and surprises need to be incorporated in the concepts and measurements of fragility, vulnerability and resilience, to make them meaningful in a practical context. It is not sufficient to use frequency distributions to describe these concepts as in the figures above.

ANTIFRAGILITY

We will now consider the antifragility concept and reflect on how to measure the degree of antifragility. This concept is more complex than the other three discussed above, as it involves a development over time: by embracing randomness, variation and uncertainties, system performance will be improved, and we are able to reach the ideal state as described by Figure 5.5. The idea is well-known and a fundamental principle in physical training. To become good, you have to impose stressors.

This idea means that we need to think of both positive and negative consequences C. Vulnerability and fragility focus on the negative consequences. Let the stresses be A_1, A_2, ..., A_n, on the system, and the associated consequences C_1, C_2, ..., C_n. These consequences will typically be minor, and the more antifragile, the higher the probability of large positive consequences and the lower the probability of negative consequences. Using the (C,U) nomenclature, a high level of antifragility is seen if judgements of (C',Q,K) for future activities are high for positive consequences C' and low for negative C'. The triplet (C',Q,K) can be viewed as a description of risk (see Section 2.3). Making such judgements over time will reveal a possible trend towards antifragility, the ideal state. The K will include information concerning past events A_i and consequences C_i.

No non-trivial system can be fully antifragile; it can only be antifragile to some degree. Its measurement depends on judgements such as (C',Q,K). This is in contrast to the antifragile ideal state as shown in Figure 5.5. As with fragility, vulnerability, robustness and resilience, we need to incorporate uncertainties and surprises in the antifragility concept and its measurement to make it meaningful in a practical setting. Two examples will be used to explain this in more detail. The first relates to the case of John the speaker introduced in Section 1.2, while the second addresses the operation of an offshore oil and gas installation as described in Section 1.3.

EXAMPLE 5.1: PERSON CONDUCTING TALKS

To become a good (brilliant) speaker, the person – John – trains a lot. He performs a number of trial-talks as well as real ones, with different types of audiences. In this way John allows stresses A_1, A_2, ..., A_n, to be activated, the results being C_1, C_2, ..., C_n, respectively. John experiences variations in outcomes. Normally the response is very good and he is pleased with his own performance. At other times the results are more average; the audience is following him with interest, but there is nothing extraordinary in his performance that makes people remember the talk. Sometimes, although very rarely, the outcome is even worse; he lacks enthusiasm and his contact with the audience becomes gradually poorer. However, he has avoided 'catastrophes' or 'disasters' – for example, where many people leave or he is completely embarrassed. He always thinks about ways of improving his talks, and following each performance he carries out a review and thinks of ways of making improvements. He sees the trial-talks as a key instrument for progress. As a speaker, he is to a large extent antifragile.

'To a large extent' is the right wording, since a 'disaster' could happen next week. John is striving all the time for new and better approaches and to achieve improvements, but this involves risk. Although his performance has been thoroughly tested at various trial-talks, an idea could have unforeseen consequences. For example, he might work on being more present, passionate and funny on the stage, with the risk that words may be said that hurt or upset some people in the audience.

At a specific point in time, we can make a judgement about the level of antifragility. The basis is all the earlier performances, both trial and real, with the stresses imposed and the consequences with respect to both positive and negative outcomes, but also judgements about the robustness of his approach, as well as his ability to use his experiences and insights to learn and make improvements. Based on his knowledge, John can make a judgement about the uncertainties and likelihood for future talks. For example, he may judge that it is very likely (much higher than 50%) that he will get an overall positive score for the coming performance, and that it is also quite likely that he will obtain a top score (a brilliant talk). His knowledge supporting these judgements is strong, but there is also a risk element linked to negative consequences that should not be overlooked, as discussed above. A 'disaster' could happen, and risk management is also about measures to manage this risk. Focusing on robustness and resilience is obviously important, so are risk assessments and their use. These are not about assigning a precise and correct probability for all types of extremely unlikely scenarios, but to:

(a) improve the overall system knowledge;
(b) identify failure scenarios, and signals and warnings;
(c) identify and evaluate measures to deal with these.

The aim must be to improve the overall understanding and treatment of risk, and in particular the awareness of and sensitivity to the details that are important for obtaining a high level of performance and avoiding disasters, in line with the concept of (collective) mindfulness.

These are the types of issues that risk assessment and management should address. Clearly, these are not the same as those rejected by Taleb: for example, striving for accurate probability and risk estimates. For dealing with rare events, unforeseen events and the surprises, searching for such estimates is obviously not an adequate approach. We need a different risk perspective and, in line with the (C,U|A,S) thinking, risk is (C,U) (or (A,C,U)), and the corresponding risk description covers (C',Q,K), i.e. unconditional vulnerability. As mentioned before, the concept of (collective) mindfulness can be nicely rooted in this way of looking at risk, since it allows for and focuses on the risk sources: signals and warnings, failures and deviations, uncertainties, probabilities, knowledge and surprises (Aven and Krohn 2014, Khorsandi and Aven 2013). The concept of collective mindfulness can help us to see these attributes and take adequate actions.

Risk has a focus on undesirable consequences, whereas the antifragility concept also relates to desirable and positive consequences. However, the (C,U) framework is general and allows for both positive and negative consequences.

Taleb (2012) refers to a test to detect antifragility (and fragility) using asymmetry: "Anything that has more upside than downside from random events (or certain shocks) is antifragile, the reverse is fragile". In the case of John's talk, there is asymmetry: more random events and situations lead to improvements than the reverse. A long list of such events can be made; see Table 5.3, which shows a few examples. However, these events are all historical; the future is not given by the past. There are risks and uncertainties present, as expressed by the risk concept (C,U) and its measurements, as discussed above. Risk management is about measures trying to deal with these risks. This includes a focus on robustness and improvements, but also the event and scenario part, through activities such as (a) to (c) mentioned above.

Taleb (2012, pp. 268–271) also presents another test for fragility and antifragility: for the fragile, shocks bring higher harm as their intensity

Table 5.3 Examples of events in the talk case and associated judgements of upside and downside (based on Aven 2014b)

Events	Effects (upside, downside)	Overall judgement + *upside dominating* – *downside dominating*
Several difficult questions from a person	The speaker was able to give meaningful answers in a convincing form	+
Audience having a different background than expected	The speaker was able to adapt to the new situation by restructuring the talk and highlighting issues that were suited to this audience	+
The speaker was very tired	The speaker was not able to show the normal enthusiasm The speaker made plans for how to avoid a reoccurrence of this situation	–
The speaker was disturbed by a nasty email from a colleague just before his/her performance	The speaker made plans to avoid exposure to such information so close to the performance At the beginning, the speaker was strongly affected by the email, but soon he/she had forgotten it and was better than ever	+

increases (up to a limit), and for the antifragile, shocks bring more benefit (or less harm) as their intensity increases (up to a point). In the talk case, if the training is intensified with more frequent and tougher trial talks, and if John is antifragile, he will experience improved results. If the intensity is doubled, his performance is perhaps four times better, measured on some reasonable scale. For the antifragile, there is a stronger than linear relationship between shock intensity and performance. This obviously works up to a point; if the intensity were to be very high, he could be burned out, becoming extremely exhausted and losing confidence. It is immediately clear from the examples in Table 5.3 that they are to be seen as nothing more than crude indicators of the level of antifragility. We will return to these tests in the discussion following the offshore case.

EXAMPLE 5.2: OFFSHORE INSTALLATION

We refer to the example introduced in Section 1.3. The main goals for the operator of the installation are to maximise values and avoid severe incidents, including accidents. A lot of minor events occur, for example gas leakages, and the production capacities also vary due to different operational measures, in particular maintenance activities. These events and activities can be viewed as the stresses $A_1, A_2, ..., A_n$, and we refer to the associated consequences as $C_1, C_2, ..., C_n$ as before. The observed variation in outcomes has been small, because the initiating events (leakages) have not escalated to accidents. However, there is a possibility that the next leakage could result in a major accident, such as the Piper Alpha disaster in 1988 and the *Deepwater Horizon* disaster in 2010. Fire and explosion scenarios may happen, leading to loss of lives and environmental damage as well as economic loss. In an antifragile system such disasters do not occur. However, as we have stressed repeatedly, for any real-life system, we cannot ignore the possibility of a major negative event occurring. The issue is then how we can analyse the degree to which the system is antifragile and possible trends, and, more importantly, how to make the system more antifragile.

First, let us reflect on the concepts of fragility and vulnerability. How fragile is the installation? A number of vulnerability analyses are carried out for such installations (Vinnem 2007), for example when studying the effect of gas leakages. These analyses, which are input to the risk analyses, describe potential scenarios starting from the leakage that can lead to a major accident, with an associated assessment of probability and uncertainties. From such analyses we can make judgements about how easily the system could break, i.e. its fragility and its vulnerability. Several gas leakages occur every year on the installation, but if the computed probability for a major accident is relatively low, it seems reasonable to conclude that the fragility is quite small and the vulnerability is low.

The assessments are based on a number of assumptions and simplifications. Many scenarios (starting with a leakage) may be excluded because they have not been thought of by the analysis group, or they may be ignored because the probability is judged to be very low. Fragility and vulnerability also relate to these events. There is clearly no objective description of the fragility level; it is strongly dependent on the analysis carried out, with its methods and the analysts involved.

As with the previous example, we can make a judgement of the level of antifragility at any time by looking at historical performance, as well as using judgements about the robustness of the system (as defined above based on $(C',Q,K|A,S)$) and its ability to learn and make improvements. The robustness of the system is studied in risk assessments, typically by means of event trees and physical modelling of the phenomena involved – for example, gas dispersion and fire development. The physical phenomena are well understood, and we can ignore strict unknown unknowns, i.e. events occurring that are completely unknown to the scientific environment. However, black swan events as mentioned in Sections 1.7 and 3.4 – events that were not on the list of known events from the perspective of those who carried out the risk analysis (unknown knowns), or events that were on the list of known events in the risk analysis but which were found to represent a negligible probability – could still occur. The level of antifragility must in some sense reflect the risk related to such events. In an antifragile system these events do not occur, but in real life we cannot exclude them, and any measure of the level of antifragility must address this risk. In addition, judgements need to be performed about the system's ability to learn and make improvements. Here ideas such as those mentioned in the previous example should be highlighted, to improve the overall understanding of risk and in particular the awareness of and sensitivity to the details which are important for achieving a high level of performance and avoiding disasters; this is linked to the concept of (collective) mindfulness. Judgements of issues like these provide a way of measuring how the system and its personnel are able to deal with events and situations occurring, and also how they are able to learn and improve.

Principles and means other than these factors related to collective mindfulness can also be used for such a purpose – for example, features from the quality discourse as discussed above. Based on the (C,U) risk perspective, and integrating the ideas from collective mindfulness and the quality discourse, a new way of thinking about risk may be developed. This thinking can be viewed as a means for obtaining a more antifragile system. The above discussion concerning how to measure antifragility also provides input regarding how to obtain a more antifragile system. There is a huge body of risk and safety literature that gives principles and methods for how to confront this risk and these uncertainties, and the topic has been the subject of considerable research, especially on how to deal with deep uncertainties and rare events with a potential for huge consequences. See the following section.

As highlighted by Røed *et al.* (2012, p. 10), many accidents in the industry – e.g. Piper Alpha (Cullen 1990), Longford (Hopkins 2000) and *Deepwater Horizon* (Deepwater 2013), are:

> mainly a result of long event sequences, which have developed gradually for a significant period, before it comes to a 'point of no return' where control is lost, and emergency preparedness has to take over. During the significant 'build-up' period (sometimes referred to as 'spiral to disaster'), there are usually several opportunities where control might have been regained, if the awareness and understanding of the sequence of events had been sufficiently deep and thorough. The same may be observed for some of the recent precursor events in the Norwegian sector that have been extensively investigated.

This quote emphasises the importance of a proper understanding and treatment of risk, with a focus on signals and warnings, uncertainties and knowledge, robustness and resilience as some of the important features of proper risk thinking.

Let us return to the symmetry test from Taleb (2012). As in Table 5.2, some examples of random events are provided in Table 5.4 for the offshore case, with associated judgements of upside and downside. A lot of events occur where the system is stressed. In all cases up to the present an accident has been avoided, although in one situation it was close. In the talk scenario, the risk management has a special focus on situations that could lead to

Table 5.4 Examples of events (situations) in the offshore case and associated judgements of upside and downside (based on Aven* 2014b)

Events/situations	Effects (upside, downside)	Overall judgement + upside dominating – downside dominating
Technical degradation leading to a leakage	The safety barriers worked efficiently and the gas leakage was quickly stopped	+
Error during planning of an operation, too low bolt torque specified, and a gas leakage occurred	The safety barriers worked efficiently and the gas leakage was quickly stopped	+
Large gas leakage occurring during testing after a period of maintenance	The leakage was stopped after one hour. Under slightly different circumstances, there could have been a leak into the air at a significantly higher rate, leading to the build-up of a large, explosive gas cloud that would have represented significant potential for a large accident The operator implemented several measures to give better (more reliable and robust) systems in the future	–
Relatively high production rates over a long period	No increase in the number of leakages identified, from a short- or long-term perspective	+

accidents. Clearly there is risk present, and the challenge is to manage this in the best way. As in the previous example, this also includes improving the understatement of risk by identifying signals and warnings and acknowledging uncertainties and the importance of knowledge.

For the other symmetry test, it is clear that the leakage record shows quite a high number of leakages, which so far have all been stopped and further escalation avoided. In this sense the system has been shown to be robust. However, there are also signs of antifragility, as more frequent leakages (and one large one in particular) have resulted in considerable efforts aimed at reducing the leakage rate (as reported in Røed *et al.* 2012). On the other hand, some major leakage events in recent years have caused the Norwegian Petroleum Safety Authorities (PSA-N) to question whether the industry does enough to pursue improvement processes and ensure lessons are learned from previous incidents (Offshore 2011). In any case, there are risks involved – a major accident could occur – and of course the system is only antifragile to a certain extent.

DISCUSSION

We have argued that Taleb's ideas of robustness, fragility and antifragility, as visualised by Figures 5.3 to 5.5, need to be supplemented by considerations of the uncertainty dimension to give a meaningful understanding for real-life situations and to be useful in describing or measuring these concepts. The distributions shown in these figures are frequency-based and not meaningful for illustrating the risk related to future rare events. In the above analysis we have discussed issues that are important in understanding and describing such risk.

An antifragile system as defined by Taleb is exposed to some level of uncertainty and variation but is protected from adverse events. Hence it shows a direction more than a reality. Consequently, we need guidance and ways of supporting the decision-making, and adequate risk and performance concepts and assessments are needed. Above we have discussed some key issues related to these concepts, and looked at assessments to obtain such support. We have to move away from the traditional probability-based approaches towards risk conceptualisation and assessment, which focus on accurate risk modelling and estimations, and instead highlight principles and ideas linked to signals and warnings, robustness and resilience, and improvements.

With such an understanding of the antifragility concept, its main features can be summarised as follows:

(a) A system that is antifragile is exposed to stressors (uncertainties, variation and risk at rather moderate levels) to obtain improvements and high performance at a later stage. The more antifragile the system, the higher probability and 'risk' related to high performance and the lower probability and 'risk' related to low performance (such as accidents). Here 'risk' relates to the consequences addressed, the uncertainty judgements (typically using probability), and the background knowledge that these are based on, i.e. (C',Q,K). For example, high 'risk' related to future performance is obtained if it is likely that the consequences will be highly desirable, and the background knowledge is strong.

(b) To measure the degree of antifragility, 'risk' needs to be described (key elements are: consequences of various stressors, uncertainty judgements, background knowledge).

The first symmetry test indicated by Taleb, briefly outlined above in relation to the example of John's talk, can be extended and conducted in line with these ideas by considering this risk description – comparing the positive 'risk' with the negative 'risk'. A highly antifragile system is characterised by high positive 'risk' and low negative 'risk'.

Taleb's second symmetry test is related to the degree that the positive 'risk' increases with increased stressors; doubling the stressors should lead to an even higher effect on positive 'risk'. However, this test is problematic to use as discussed above; the idea can only work for a certain level of stressors, and it is often difficult to make comparisons of the levels of increase for the stressors and the risk.

Then we may ask, what does this understanding of the antifragility concept add to risk assessment and management practice compared to robustness and resilience? The key contribution is the antifragility concept's idea of linking variation, uncertainties and risk at the stress level to positive and negative "risk" at later stages. Robustness and resilience both address the stress dimension, but they do not conceptualise these in relation to future developments that extend beyond established functions. Take the example of John's talks. The issue is not only about ensuring that John is able to deal with events and surprises when he gives his talk. Equally important is the process of developing over time and reaching higher and higher levels of performance. In the offshore case the design may be difficult to change in operation, but there may be a potential for operational and organisational improvements. The antifragility concept emphasises the importance of not being satisfied with performance compliance at specific points in time. What is coming next must always be highlighted. If training programmes are implemented with the purpose of increasing the understanding of risk in operational teams carrying out critical operations, such as well operations, the result may be a temporary increase in accident risk as procedural compliance thinking is challenged, but a considerably reduced risk over time. In the design stage of an installation, the use of different types of equipment and arrangements for meeting specific functions can lead to variation and more or less strong performance, but the resulting different experiences may be decisive for the development of next-generation units with considerably better performance and reduced negative risks.

No one can disagree with Taleb when he says: "It is far easier to figure out if something is fragile than to predict the occurrence of an event that may harm" (Taleb 2012, pp. 4–5). However, fragility is also often difficult to measure. As with vulnerability, fragility can capture situations with a poor knowledge basis and where future performance is subject to large uncertainties. In this chapter we have shown how the concepts of vulnerability and fragility are closely linked to risk, when suitably interpreted. As with risk, measuring vulnerability and fragility is also dependent on judgements, and important contributors can be lost in the representations. Black swan events are also present at the vulnerability and fragility level as surprising sequences of events and conditions. Taleb states that risk is not measurable for real-life situations, in the sense of accurate estimations of an objective true value. However, risk can be described, and that is where the benefit of risk assessment lies. Such descriptions capture much more than the assigned probability numbers of rare events. Clearly, in the case of black swan events it is not the probability numbers for the events that are of interest but the proper understanding of risk, the signals and warnings, the awareness and sensitivity to operations, resilience and so on, as thoroughly discussed in this book. Taleb proposes "to stand our current approaches to prediction, prognostication, and risk management on their heads" (Taleb 2012, pp. 4–5), and in the context of much of the current thinking with its focus on probability modelling and estimations, Taleb's view is understandable. However, as argued in this book, risk management is required in order to find the proper measures to confront potential events that may occur. There are always limited resources available for this purpose, and risk assessment provides decision support. Decision-makers need to be informed about the issues discussed above related to precursors, uncertainties, the collective mindfulness principle, and so on. In a particular case, a decision-maker may need to choose between investment in measures that are effective in the case of some events but not others, and investments in some other measures with the opposite effect. Accurate predictions and estimates cannot be provided, but, certainly in most cases, informative risk descriptions can.

CONCLUSIONS

To be able to see how Nassim Taleb's antifragility concept can be used in a risk management context, we need to understand what this concept expresses in relation to common ideas and principles of risk assessment and management. The above analysis has contributed to this end by using the general risk set-up of Sections 2.3 and 2.4, and by providing a meaningful interpretation of the antifragility concept. It is argued that Taleb's antifragility concept as described by him needs to be supplemented by considerations of the uncertainty and risk dimension. By doing this, the concept adds an important contribution to the current practice of risk assessment and management with its focus on the dynamic aspects of risk and performance, seeing variation, uncertainties and risk as the stress level in relation to positive and negative 'risk' linked to future consequences, as discussed above.

5.2.4 Use of the extended risk and performance perspective

The main uses of the perspective are:

(1) As a general guideline for designing methods and tools for understanding, assessing, managing and communicating risk and safety.
(2) As a means for evaluating and suggesting improvements for the quality ("goodness") of various risk management activities, such as the use of various types of risk assessment, barrier principles and risk management strategies – for example, the use of the ALARP (As Low As Reasonably Practicable) principle.
(3) As a supplement to standard accident investigation procedures, by drawing attention to critical issues such as the violation of the five collective mindfulness principles, and issues related to knowledge,

uncertainties, variation, the unforeseen, surprises and black swans, from different perspectives (individuals and groups) and points in time. See UCB (2010, p. 82) and Aven *et al.* (2014b).

(4) As a means for evaluating the quality ('goodness') of a concrete risk assessment being conducted; see Section 4.4.

(5) As a means for evaluating the quality ('goodness') of the risk understanding in relation to critical operations, reflecting the ability to understand the total system, relevant knowledge, transfer of experience and learning; see Section 4.4.

To illustrate the use of the perspective, let us look at some examples from the petroleum industry. First, consider the suggested guidelines from the Norwegian Petroleum Safety Authority (PSA-N) for barrier management principles (PSA-N 2013a). With respect to concepts, principles and the main thinking behind the understanding, assessment and management of risk and safety, it can be concluded that the document is very much in line with the integrated framework presented here. For example, this applies to the meaning of the concept of risk. A further analysis of the five collective mindfulness principles and aspects highlighted by the quality discourse reveals two main challenges. The first one relates to the *reluctance to simplify* principle of the collective mindfulness concept. Although the barrier principles highlight the need for a total system view with a focus on total barrier functions, the strong emphasis on the specification of detailed barrier element performance requirements may lead to difficulties in practice. Meeting the barrier element performance requirements may give the false perception that the risks are low and the barrier functions are fulfilled. As we know, the connections between barrier element performance, risk and satisfying barrier functions are often unclear. Good performance numbers for the detailed barrier elements are no guarantee for safety. Holistic thinking is important, particularly for being able to 'identify' black swan events as well as for ensuring robustness and resilience. From a practical point of view, we may find that both the industry and the agency, through their auditing, are happy with a regime that highlights the barrier element because it is simple and easily followed up compared to broader, more judgemental assessments of barrier function performance and risks. The PSA-N is aware of this challenge, and it will be interesting to see how the implementation process will proceed.

The second challenge of these barrier management principles relates to management by objectives and the compliance focus, and is linked to the discussion above concerning the reluctance to simplify. The approach recommended by the PSA-N document has a strong emphasis on formulating, assigning and satisfying performance requirements, which can easily lead to an overly strong focus on meeting requirements rather than on identifying the best solutions and measures overall. Experience has shown that such an approach represents a serious challenge – the compliance regime prevails over the improvement processes, which are always highlighted in theory but often fail to be given priority in competition with the convenience and practical attractiveness of compliance procedures. See the discussion in Section 4.2 (adaptive analysis).

5.3 How to deal with deep/severe uncertainties?

A main foundational issue in risk assessment and risk management is how to handle deep uncertainties in relation to the events occurring and the consequences of these events, such as in preparing for climate change and managing emerging diseases: what policies and decision-making schemes should be implemented in such cases? Traditional statistical methods and tools are not suitable, as relevant supporting models cannot easily be justified and necessary data are missing. However, other approaches and methods exist, and here we are particularly interested in methods for robust and adaptive risk analysis. In a recent paper, Cox (2012) provides a thorough analysis of using such methods to meet deep uncertainties. He argues that these methods provide genuine breakthroughs for improving predictions and decisions when the correct model is highly uncertain. He reviews ten tools that can help us to better understand deep uncertainty and make decisions even when correct models are unknown: (subjective) expected utility theory; multiple priors, models or scenarios, robust control, robust decisions; robust optimisation; average models; resampling; adaptive boosting; Bayesian model averaging; low regret online detection; reinforcement learning; and model-free reinforcement learning. These tools are based on two strategies: finding robust decisions that work acceptably well for many models (those in the uncertainty set); and adaptive risk management, or learning what to do by well-designed and analysed trial and error (Cox 2012). Adaptive risk management is a technique which seeks to treat risk by considering a set of alternatives and dynamically tracking these in order to gain relevant information and knowledge about the effects of different courses of action; see Section 4.3. Cox's analysis is based on a specific perspective on uncertainties: those things that distinguish deep uncertainties "from the more tractable uncertainties encountered in statistics and scenario analysis with known probabilities" (Cox 2012, p. 1608). Cox refers to a specific uncertainty taxonomy presented by Walker *et al.* (2003) (and partly based on the ideas of Courtney [2001] and Walker *et al.* [2003]). This taxonomy is discussed in Aven (2013f) and will not be discussed further here. It relates to the meaning of deep uncertainties which we have examined closely in Section 3.3; see also Section 5.1.

Instead, we shall focus on the managerial issues. A risk assessment informs the decision-maker, but it does not prescribe what to do as the decision-maker has to consider aspects that go beyond the results of the risk assessment. The benefits related to the activity studied, as well as strategic and political concerns that could be important for the decision, may not be captured by the assessment. The decision-maker also has to take into account that the assessments are based on many assumptions, and that they in turn have limitations. Such considerations are often referred to as risk evaluation or managerial review and judgement; see the discussion in Section 3.3. They represent the leap between the analysis part and the decision.

Sometimes attempts are made to reduce this leap by introducing decision-making procedures such as the use of risk acceptance criteria or limits, see discussion in Section 5.1. If the risk assessment shows a probability number below a criterion x_0 the risk is considered acceptable, whereas if the probability exceeds x_0 the risk is considered unacceptable.

In practical risk management there is always a search for procedures like this to simplify the decision-making processes. Nonetheless, there will be a need for careful review of the approach taken since it will have limitations and weaknesses. One obvious problem is that risk is not adequately reflected by the probability numbers alone, especially if the uncertainties are large, as described here. The point is that the analytical framework used by the analysts and experts could be based on assumptions that turn out to be wrong or do not adequately reflect all aspects of the situation, and the decision-maker always needs to take this into account. Hence a managerial review and judgement is required that can see beyond the narrow technical criteria when making judgements about whether or not the risk is acceptable.

This is also the case when decision analyses (including cost–benefit analyses [CBAs]) are used. Perhaps the situation does not seem so obvious in this case, since the decision analysis is designed to reflect the decision-maker's attitude to risk and cover all the aspects that are important for decision-making. Following the decision analysis theory, we should choose the alternative that maximises/minimises the expected utility, reflecting the performance of the alternatives considered, the values and uncertainties. However, here it is also necessary to review the decision-making basis: what is the background knowledge of the analyses? What assumptions and suppositions are made? Which decision-making options are being analysed? Who are the experts assigning the probabilities? How is the utility function derived? To what degree does the utility function reflect the decision-maker's preferences? And so on. We are led to a managerial review and judgement. It means that the analysis is strictly an aid for decision-making. This is comparable to the 'moderate' decision analysis view, as mentioned by Fischhoff *et al.* (1981) and supported by a number of decision theorists (e.g. Watson and Buede 1987, French and Insua 2000). The decision-maker needs to place the results of the decision and risk analysis into a larger context of review and judgement. See Section 5.1.

Walker *et al.* (2003) and Cox (2012) refer to weights on outcomes, such as value functions and utilities, and the uncertainties of these weights. The meaning of such uncertainties is not clear. Let us simplify and say that C expresses the quantity of interest, for example the:

- Number of fatalities N.
- A frequentist probability p that N exceeds 100 (here, a probability model is introduced).

In the risk assessment, uncertainties are described and probabilities specified but normally value functions or utilities u(C) are not. However, let us think of a case where a utility function is also assigned. Following the subjective expected utility theory, this utility function reflects the preferences of the decision-maker. According to this theory, there is no uncertainty in this function because there is no underlying correct value. Maybe the decision-maker feels some practical difficulties in assigning the function values, but we should be careful in using a term like 'uncertainty' for this; rather, we should speak about 'imprecision' or 'vagueness'.

Alternatively, is the referred uncertainty about the utility function to represent the uncertainty that some analysts have about which utility function the decision-maker has or will have at a future point in time? Conceptually such uncertainties are meaningful, but they would be in conflict with the basic ideas of the expected utility theory.

Cox (2012) refers to several obstacles in the way of using the subjective expected utility (SEU) theory, including uncertainty about the full range of consequences, uncertainties about the probabilities and uncertainties about the values and preferences (utilities). The first problem relates to unknown unknowns and black swans, and is obviously a problem for any formal approach to guiding decision-making, not only SEU. The second obstacle relates to the meaning of a probability. SEU is based on subjective probabilities and there is no reference to correct probabilities. Such probabilities do not exist. However, Cox is probably referring to the notion that probability models are then justified and there is consensus about the model. As regards obstacle three, Cox refers to differences in views concerning, for example, the willingness to take on risk to achieve some potential rewards, but also because future preferences are not known today – i.e. two completely different aspects of uncertainty or differences in values between people. Mixing these concerns may lead to confusion, as the risk management will be strongly dependent on the subject of concern.

In most cases there are many stakeholders and the decision is not made by one person. In theory we can produce utility functions for each stakeholder. Are the uncertainties perhaps related to the utility function expressing the variation in the utility function among these stakeholders?

Think again about the decision problem about whether or not to open the Barents Sea and the Lofoten archipelago in Norway for year-round petroleum activities in especially vulnerable areas (Aven and Renn 2012). The issue is not so much about deep uncertainties but about different views on the weights to be given to the benefits in comparison with the environmental values at stake – in other words, the differences in the willingness to take on risk to achieve some potential rewards, as Cox mentions. But is the issue of robustness concerned with making decisions that are 'relatively good' for all such views? No, robustness as studied in risk analysis would normally be kept outside such considerations (as they are political), unless one has the specific goal of seeking solutions that are 'good' for a set of views as in a coalition of parties.

Clearly, there is a need for robustness when facing deep uncertainties (in the sense that accurate prediction models cannot be established), because the outcomes are difficult to predict. An emergency preparedness system is attractive if it is able to work efficiently for many types of undesirable events that may occur, since it is difficult to foresee exactly which scenario will happen in the future. However, the fact that some political parties strongly dislike uncertainties in relation to environmental damage, whereas others find the uncertainties small and acceptable given the benefits of the activity, should not necessarily drive solutions that are robust against both views. Within a coalition such robustness may be sought, but in general care has to be shown when allowing for variation in the utility function to form the basis for the robust solutions.

In his conclusions, Cox (2012) argues for a shift in thinking about risk analysis, going from fundamental questions about the risk – what can go wrong, what are the consequences and the likelihoods? – to questions like: is there clearly a better risk management policy than the one I am now using? How probable should I make each of my possible actions? Would a different choice of policy give me a lower level of regret (higher expected utility of consequences) given my uncertainties? The focus should change from the risk description to a consideration of how I should act now and in the future.

The motivation for this recommendation is not difficult to follow, since what matters is not the risk description but what we do about the risk. However, the comparison should not be between the risk description and decision-making issues, but between risk management based on risk descriptions and managerial review and judgement on the one hand, and the decision-making issues on the other. Then, a fairer comparison may be made, and the conclusions are not so obvious. We all find robust decision-making tools useful in providing decision support, but their role is of less importance in a regime where managerial review and judgement is acknowledged, as recommended in this book, compared to the thinking that seems to be the basis of the reasoning among many decision analysts. Figure 5.6 illustrates the point made here. The upper row refers to a prescriptive use of decision analysis, as recommended for example by Lindley (1985), while the third row illustrates an analogous stand with the robust decision analysis as the basic framework. The second row shows a typical risk perspective approach to decision-making, while the bottom row illustrates the approach argued for in this book, where all the tools are considered as decision support tools to be followed by a managerial review and judgement.

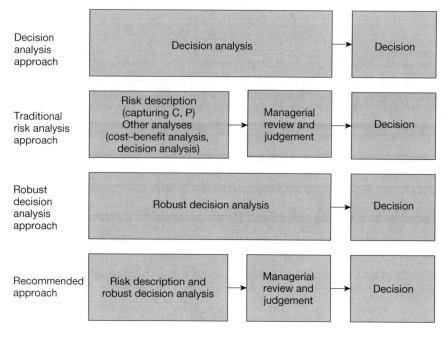

Figure 5.6 Four approaches to linking decisions and decision analysis (based on Aven 2013f)

The need for managerial review and judgement is acknowledged in applications to a varying degree. In industry, for example, the use of mechanical procedures for making judgements about risk acceptability based on calculated risk numbers is common (Aven and Vinnem 2007).

Above we have discussed the issue of managerial review and judgement in relation to robust optimisation and other approaches for confronting deep uncertainties. Although the point of departure for the discussion is Cox's work (2012), it should be noted that this specific work does not express any view on the need for a managerial review and judgement.

CONCLUSIONS

The idea that decisions in the context of uncertainties should be completely specified by the result of a decision analysis has been common among scholars of the economic decision-making school and among Bayesian decision-making theorists (e.g. Lindley 1985). However, the current thinking is more that these analyses provide decision support: there is a need to see beyond the analyses to make effective decisions. However, there is no consensus on what 'going beyond' should capture. Some authors see an important principle gap between the analyses and the decision-making that can never be bridged, whereas others consider the 'beyond' aspect to be more of a technical issue that can be solved by improved analysis methods. Many decision analysts seem to belong to the latter category. This book has argued for the former perspective as a fundamental description of the decision-making process, in general and in relation to deep uncertainties in particular, and refers to this 'beyond' as managerial review and judgement. Key aspects of review and judgement have been identified. When it comes to handling deep uncertainties, standard probability-based tools are inadequate. It is a research challenge to develop suitable approaches and methods for this case. The present book provides reflections on some of these approaches, as discussed by Cox (2012) and others.

5.4 Discussion and conclusions. How to confront the possibility of surprises (black swans)?

The above analysis has shown how the perspectives on risk affect the way risk is assessed and described in practice. The changes seen are related inter alia to the knowledge dimension and potential surprises (black swans). Assigned probabilities and expected values still represent an important tool to reflect information, uncertainties and degrees of belief. However, the strength of knowledge on which the probability-based figures are based is an essential element of the risk description and needs to be highlighted. The best way to do this is a research topic in itself; there is a need to explore many different approaches. In this book we have looked at some methods, using simple qualitative categorisations to reflect this strength. These methods are partly built on known tools but include several novel features, including the 'assumption deviation risk' concept. In addition we have outlined procedures for describing risk related to black swans, i.e. surprises compared to the beliefs of the experts and analysts involved in the risk assessment. The change in risk description affects risk management and decision-making in different ways. Probability-based procedures need to be adjusted, adequately reflecting the knowledge and surprise dimensions. Examples are provided for how to modify the use of risk acceptance criteria and associated ALARP processes.

To confront the possibility of surprising events and black swans, we need to balance risk-based approaches, cautionary/precautionary- (robustness, resilience, adaptive) and discourse-based approaches. This is the general answer and is fundamental for risk management, with or without a special focus on surprises and black swans. Only in cases when the knowledge is very strong and the uncertainties small can the risk-based approach alone be used. In most situations, all three strategies are required. The challenge is to find a satisfactory balance between these approaches and strategies. When the stakes are high and the uncertainties large, we obviously need to highlight robust and resilient solutions and arrangements in order to be prepared in case some extreme unforeseen events should occur. Potential surprises and black swans call for robustness, resilience and antifragility, as discussed above.

In the following some reflections will be provided for each of the three categories of black swans defined in Section 3.4:

(a) Events that were completely unknown to the scientific environment (unknown unknowns).
(b) Events not on the list of known events from the perspective of those who carried out a risk analysis (or another stakeholder), but known to others (unknown knowns – unknown events to some, known to others).
(c) Events on the list of known events in the risk analysis but judged to have negligible probability of occurrence, and thus not believed to occur.

5.4.1 Black swans of the unknown unknown type

Unknown unknowns are events that were completely unknown to the scientific environment, and it is of course difficult to be prepared for such events. Focusing on resilience, signals and warnings provides useful general means in addition to scientific work generating knowledge about the relevant phenomena. In general, increased knowledge reduces the probability of a black swan of this type. Hence, testing and research are generic measures to meet this type of black swan risk.

Take the swine flu outbreak in 2009 as an example. It was caused by a type A influenza (H1N1) virus. The World Health Organization (WHO) declared that the flu had developed into a full-scale world epidemic, and a vaccine was quickly developed. In some countries (Sweden, Finland, Norway and Iceland), the authorities explicitly set the goal of vaccinating the whole population. The illness turned out to be quite mild, but it had some severe side effects that were previously unknown; see e.g. Munsterhjelm-Ahumada (2012). These side effects came as surprises – they were black swans of the unknown unknown type.

The vaccination was carried out because the authorities believed that the flu itself would cause serious illness and problems, at a much higher level than the side effects. Normally there is time for fairly thorough testing of vaccines to control the risk related to side-effects, but in 2009 this was not the case. The uncertainties were large.

Obviously, there could be unknown side effects in the case of vaccination, and analyses and judgements need to be conducted to characterise the risks. The problem here was that the decision concerning vaccination had to be taken very quickly. It was impossible to avoid a weak knowledge base. There was no time for thorough testing and research and adaptive management. The authorities also had to balance the need for faithful risk characterisations and the desire to get the population vaccinated. In the Nordic countries mentioned above, the authorities initiated public relations campaigns which could be described as 'moral persuasion'. Solidarity became the slogan: be vaccinated to protect your fellow citizens (Munsterhjelm-Ahumada 2012).

Faithful risk characterisations addressing possible unknown side effects (black swans) were not very well highlighted; the black swan risk was not really an issue. One might speculate about whether this was a deliberate policy. There is no doubt that the decision was a difficult one for the authorities because of the time pressure; they had to balance difficult judgements about the development of the flu, the efficiency of the vaccination, risk and uncertainty issues, as well as ethical aspects.

On this basis it is not surprising that so many people decided to take the vaccine. The decision became quite easy, following advice from the authorities.

From the individual person's point of view, one could argue that the black swan risk should have been reported more faithfully. To make an adequate decision, one has to be risk-informed. However, it is not straightforward to decide how such information could best have been communicated. The knowledge base is weak, and it is impossible to express meaningful numbers characterising risk. Instead, we have to rely on more general qualitative statements. Here is a suggestion for how the risk could have been described and communicated:

> This vaccine could have unknown side effects. There are uncertainties. We think it is unlikely that severe side effects will occur, but the knowledge base is rather weak and we cannot exclude the possibility.

It is not enough to limit the statements to probability characterisations. Expressing that it is unlikely that severe side effects will occur without also referring to the knowledge base and the potential for surprises would mislead the receiver. The knowledge supporting the probability is as important as the probability itself.

5.4.2 Black swans of the unknown known type

A black swan of this type is an event that is not on the list of those identified by the relevant risk assessment, but it is not an unknown type of event. Its possible occurrence is known by other persons, groups or communities. We can formalise it in this way: A' are those events that we have identified in the risk assessment, and A is the occurrence of the actual event, which is a type of event known by others than those involved in the risk assessment. The event is a black swan (or a near black swan) of the unknown known type if A is not covered by A'.

To meet this black swan type of risk we need:

- Improved risk assessment to identify these events.
- Improved communication to transfer knowledge to relevant persons.

We discussed these issues in Chapter 4. The key is knowledge building and the transfer of experience and learning, since the knowledge about A is available but is not possessed by the relevant analysis and/or decision team.

5.4.3 Black swans of the probability type judged negligible

This third type of black swan event constitutes those that are on the list of known events in the risk analysis but whose probability of occurrence is judged to be negligible, and thus they are not believed to occur. Yet such events do happen. How should we deal with this type of event? Should we just accept the risk? We have accepted that there is a small probability of an extreme event occurring, meaning that the event could occur.

Recall the discussion in Section 3.4 about this issue, relating to perfect storms. The knowledge base is weak, and the probabilities are subjective (judgemental, knowledge-based), and may be more or less strongly founded. Hence, it is appropriate to scrutinise both the judgements about acceptable risk and negligible probability, and the background knowledge that supports these judgements.

Such a scrutiny needs to be based on the acknowledgement that:

(i)   acceptable risk should not be determined by judgements about probability alone.
(ii) events may occur even if very low probabilities are assigned.
(iii) the cautionary and precautionary principles constitute essential pillars of the risk management linked to such events (black swans).

The risk analyst may derive a set of probabilities for specific events to occur and combine them with different loss categories, but these numbers must be seen in relation to the strength of knowledge that supports the probabilities. We may have two situations with the same probabilities, one where the assignment is supported by a strong evidence base and one which relies on very poor background knowledge, as we have pointed out many times before (see for example Section 2.2.2). In engineering contexts, common practice is built on probabilistic criteria (such as a $1 \cdot 10^{-4}$ probability limit) to determine what is an acceptable design (Section 5.1). Such an approach cannot in general be justified because it ignores the degree of knowledge that supports the probability assignments. The assignments may be based on many critical assumptions, and these assumptions might conceal important aspects of risk and uncertainty. For instance, we may assume that the present system is a standard one, but it could turn out to have special features, for example being extremely sensitive to specific hazards.

An adjusted procedure reflecting the strength of knowledge supporting the probability assignments is presented and discussed in Section 5.1. The approach relies on cautionary thinking. It generates a process that searches for measures to reduce risk and avoid the event occurring, despite the fact that the judged probability is very low.

The cautionary and precautionary principles are important risk management principles and were thoroughly discussed in Section 5.1. Think about the vaccine example introduced in Section 5.4.1. Here we faced scientific uncertainties concerning side effects, and if a person did not take the vaccine he or she could refer to the precautionary principle. Health experts might point out that the probability of side effects is low and hence acceptable, but the strength of knowledge supporting this type of statement is poor and there is a need for considerations that give due weight to the uncertainties as discussed above.

A number of safety and security measures are justified by reference to the cautionary principle. We implement robust design solutions to be able to meet deviations from normal conditions, we implement emergency preparedness measures even if the probability of their use is very small, and so on. We see beyond the probabilities because we know that surprises can occur relative to our judgements. This is being cautious.

Consider the Fukushima Daiichi nuclear disaster in Japan in March 2011. Here the risk was accepted. The probability that such an event would occur was considered to be so low that the risk was judged acceptable. The judgement was based on many considerations and assumptions as discussed by Paté-Cornell (2012). These considerations and assumptions can obviously be questioned because earthquakes from the ninth and seventeenth centuries caused tsunamis reaching heights far beyond the design criterion of the plant, and these were not accounted for in the design of the nuclear reactors.

It is not obvious that better risk assessment would have led to vital changes in the Fukushima case, but it could have. Several weaknesses in the Fukushima assessments were indicated by Paté-Cornell (2012), and there is also the potential for general improvements in risk assessments, as discussed in Chapter 4, by giving further attention to the knowledge and surprise dimensions.

5.4.4 Final remarks

There are certain aspects to which we need to pay more attention in comparison with typical current practices, and the book has pointed to and highlighted some measures linked to the way we should think in relation to these issues. The main thesis of the book is that by adopting the new risk perspectives, with their additional features as described in Section 5.2, there is the potential for improved understanding and better assessment and management of risk, surprises and black swans. The proposed perspective can help analysts and decision-makers in two main ways:

(1) It provides appropriate concepts and a platform for a deeper under-standing of what the risk associated with surprises and black swans is all about.

(2) It provides analysis and management principles that can prevent, or at least reduce the probability of, black swan events (which have negative consequences), and in addition can stimulate and lay a basis for the development of appropriate specific methods that can achieve such an effect.

In risk analysis, events and scenarios are identified, for example hydrocarbon leaks, and barrier systems are put in place in case such events should in fact occur. Many types of events happen in the course of a year, but they do not have serious implications because the barriers work as intended. This is also the case for so-called near misses – but the margins are in some cases small. Minor changes could have resulted in a disaster. When a major accident occurs, it is often because there are several 'surprising events' rather than just one. To address these issues, it is important to have an understanding of the various concepts and how they relate. This book contributes to this understanding (as indicated in point 1 above).

In addition, the book can provide specific help on how we should proceed in order to improve the understanding of risk and better analyse and manage risk including a consideration of black swan events. Knowledge and uncertainty are key concepts. Black swans are surprises in relation to someone's knowledge and beliefs. In the September 11 example some people had relevant knowledge, while others did not. In the Fukushima example it was the judgements and probabilities that were essential, but they were based on data, information and arguments/opinions, so here too the issue is knowledge. We must think beyond current practice and theory. We need new principles and methods. This book contributes to such principles and presents examples of specific methods, and also lays the foundation for the further development of a set of specific methods, thereby helping to reduce the probability of (negative) black swans.

Here is a checklist for aspects to consider in a risk assessment to ensure that due attention is given to the knowledge and surprise dimensions of risk:

Checklist

1 Is there an overview of the assumptions made (linked to system, data, models, expert reviews, etc.)?
2 Has a risk assessment of the deviations from assumptions been conducted (an assumption deviation risk assessment)?
3 Have attempts been made to reduce the risk contributions from the assumptions that have the highest deviation risk?
4 Has the goodness of the models used been assessed? Have the model errors (difference between the correct value and the model's outcome) been found acceptable?
5 Is the strength of knowledge, on which the assigned probabilities are based, assessed?

6 Is this strength included in the risk description?
7 Have attempts been made to strengthen the knowledge where it is not considered strong?
8 Have special efforts been made to uncover the black swans of the type unknown knowns?
9 Have special efforts been made to uncover any weaknesses or holes in the knowledge on which the analysis group has built their analysis?
10 Have special efforts been made to assess the validity of the judgements made where events are considered not to occur due to negligible probability?
11 Have people and expertise, who do not belong to the initial analysis group, been used to detect such conditions?
12 If the expected values of a quantity are specified, has the uncertainty of this quantity been assessed (for example, using a 90% uncertainty interval for this quantity)?

It is believed that the recommended risk thinking allows for and encourages considerations and reinterpretations of the way risk is assessed at different stages of an activity; these are essential features of a management regime supporting continuous improvements. Current risk perspectives are considered to be less satisfactory for this purpose because the frameworks presume a stronger level of stability in the processes analysed. It has been shown that the (C,U) risk perspective also provides a platform for incorporating concepts from organisation theory and learning (collective mindfulness) and resilient engineering.

Bibliographic Notes

This chapter is built on basic theory of risk management as presented for example in Aven and Vinnem (2007) and Aven and Renn (2010), as well as some recent papers in the field, mainly those of Aven (2013f, g, 2014b, c) and Aven and Krohn (2014).

The categorisation of risk problems is inspired by Renn (2008), Aven and Renn (2010) and Aven (2013f), while the deep uncertainty discussion is linked to work by Cox (2012). In relation to this discussion, it is appropriate to mention that many perspectives exist other than those studied in this chapter; for example, Karvetski and Lambert (2012) seek to turn the conventional robustness discussion away from its focus on which action is the most robust, in favour of identifying which are the uncertainties that matter most, which matter least, which present opportunities and which present threats – and why.

Section 5.2.3 on the antifragility concept is based on Aven (2014b). Although the discussion in Section 5.2.3 focuses on Taleb's antifragility concept, it also applies to some extent to his black swan concept (Taleb 2010), whose occurrence, according to Taleb, cannot be predicted, so that our only available option is to shore up the robustness and responsiveness of our society so that the shocking effects of a shocking event are not that shocking (Lindaas and Pettersen 2013). These black swan theses are studied by Lindaas and Pettersen (2013), Masys (2012) and Paté-Cornell (2012), among others, and they also provide input to the present discussion, although their perspectives differ from the one adopted here.

In addition to the literature mentioned above, examples of articles and books addressing safety and risk issues in relation to deep uncertainties and rare events are: Hollnagel *et al.* (2006), Chapman (2005), Chapman and Ferfolja (2001), Flood (1999), Kaplan *et al.* (1999), Kunreuther and Useem (2010), Reason (2004), Rochlin (1999), Van der Merwe (2008), Walker *et al.* (2013), Ranger *et al.* (2013) and Cox (2012). The improvement aspect is addressed in some of these works, but the main focus is on anticipation, robustness, resilience and similar types of measures rather than on how to become better and better, as in the antifragility concept.

Appendices

Appendix A
Uncertainty representations, with emphasis on probability

This appendix provides a review of common interpretations of probability, covering both 'objective' and 'subjective' probabilities, as well as some common alternative representations (interval probabilities, probability bound analysis, and approaches based on possibility theory and evidence theory). The work is mainly based on Aven (2013h) and Aven and Zio (2011). In the discussion, reference is made to a number of researchers on probability, but it must be stressed that this presentation does not in any way aim to provide an 'all-inclusive' review of the many persons that have contributed to the development of probabilities. The reader interested in more comprehensive historical reviews should consult, for example, Bernardo and Smith (1994), Singpurwalla (2006) and Galavotti (2000). For literature addressing the concept of probability in relation to risk analysis and safety, see Östberg (1988), Parry (1988), Kaplan (1988), Apostolakis (1988, 1990), Martz and Waller (1988), Vaurio (1990), Watson (1994, 1995), Yellman and Murray (1995), Bennett (1995) and Mosleh and Bier (1996). As these references show, the meaning of 'probabilities' was a topic of intense debate some fifteen to twenty-five years ago, mainly in relation to nuclear power risk analyses. Of course, this discussion is linked to the one raised here, but the overlap is in fact quite small, since these papers and related works do not constitute a critical discussion of the interpretation of subjective probabilities: none mentions the uncertainty standard perspective, which is highlighted in the following. In fact, there are very few texts on probability beyond the works of Lindley (2006) and Bernardo and Smith (1994) that cover this interpretation.

In the following text we will sometimes refer to the offshore platform example introduced in Section 1.3. See also Section 2.2.1.

A.1 'Objective' probabilities

This section reviews and discusses the following types of probabilities (interpretations): classical probabilities, frequentist probabilities, propensities and logical probabilities.

A.1.1 Classical probabilities

The classical interpretation, which dates back to de Laplace (1812), applies only in situations with a finite number of outcomes which are equally likely to occur. According to the classical interpretation, the probability of A is equal to the ratio between the number of outcomes resulting in A and the total number of outcomes, i.e.:

P(A) = Number of outcomes resulting in A/Total number of outcomes.

As an example, consider the throwing of a die. Here, P (the die shows one) = 1/6, since there are six possible outcomes which are equally likely to appear and only one that gives the outcome 'one'.

The requirement that each outcome must be equally likely is critical for the understanding of this interpretation. It has been subject to much discussion in the literature. A common perspective is that this requirement is met if there is no evidence that favours some outcomes over others (this is the so-called 'principle of indifference'). So classical probabilities are appropriate when the evidence, if there is any, is symmetrically balanced (Hajek 2001), as we may have when throwing a die or playing a card game.

However, this interpretation is not applicable in most real-life situations beyond random gambling and sampling, because we do not have a finite number of outcomes which are equally likely to occur. While the discussion about the indifference principle is interesting from a theoretical point of view, it is not so relevant in the context of the analysis here, where we search for a concept of probability that can be used in a wide class of applications.

A.1.2 Frequentist probabilities

A frequentist probability of an event A, denoted $P_f(A)$, is defined as the fraction of times the event A occurs if the situation considered were repeated (hypothetically) an infinite number of times. Thus, if an experiment is performed n times and the event A occurs n_A times, the $P_f(A)$ is equal to the limit of n_A/n as n goes to infinity (tacitly assuming that this is possible); i.e. the probability of the event A is the limit of the fraction of the number of times event A occurs when the number of experiments increases to infinity.

A frequentist probability, $p = P_f(A)$, is thus a mind-constructed quantity. It is a model concept, founded on the law of large numbers, stating that frequencies n_A/n converge to a limit under certain conditions. Unfortunately these conditions themselves appeal to probability – we have to assume that the probability of the event A exists, and is the same in all experiments, and that the experiments are independent. To solve this circularity problem, different approaches have been suggested. Strict frequentists attribute a concept of probability to an individual event by embedding it in an infinite class of 'similar' events having certain 'randomness' properties (Bedford and

Cooke 2001, p. 23). This leads to quite a complicated framework for understanding the concept of probability, and this reason alone could be an argument for concluding that this approach is not suitable for practical use (see also the discussion in Van Lambalgen [1990]). An alternative approach is to simply assume the existence of the probability, $P_f(A)$, and then apply the law of large numbers to give $P_f(A)$ the limiting frequentist interpretation. This is a common way of looking at probability. Starting from Kolmogorov's axiomatisation (the standard axioms for probability: non-negativity, normalisation, finite and countable additivity) and conditional probability (see e.g. Bedford and Cooke 2001, p. 40), and presuming the existence of probability, we derive the well-established theory which is presented in most text books on probability, where the law of large numbers constitutes a key theorem providing the interpretation of the probability concept. For applied probabilists, this perspective seems to be the prevailing one.

One idea for justifying the existence of the probability is the so-called propensity interpretation. It holds that probability should primarily be thought of as a physical characteristic. The probability is just a propensity of a repeatable experimental set-up which produces outcomes with limiting relative frequency $P_f(A)$ (SEP 2009). Suppose we have a coin; its physical characteristics (weight, centre of mass, etc.) are such that, when throwing the coin over and over again, the head fraction will be p.

The literature on probability shows that the existence of a propensity is controversial. However, from a conceptual point of view, the idea of a propensity should not be more difficult to grasp than the infinite repetition of experiments. The point is that if you accept the framework of the frequentist probability, i.e. that an infinite sequence of similar situations can be generated, you should also accept the propensity thesis, as it basically states that such a framework exists. Hence, for the situations of gambling and fractions in huge populations of similar items, the frequentist probability should make sense, as a model concept. If you throw a die over and over again it is obvious that the properties of the die will change, so the idea of 'similar experiments' is questionable. However, for all practical means we can carry out (in theory) a huge number of trials, say 100,000, without any physical changes in the experimental set-up, and that is what is required for the concept to be meaningful. The same is the case if we consider a population of (say) 100,000 human beings belonging to a specific category, (say) women in the age range 40 to 50 in a specific country.

Given this, it is not surprising that the frequentist probabilities are so commonly adopted in practice. A theoretical concept is introduced, often in the context of a probability model, for example the normal or Poisson distributions, and statistical analysis is carried out to estimate (and study the properties of the estimators of) the frequentist probabilities (more generally the parameters of the probability models) using well-established statistical theory. However, the type of situations that are captured by this framework is limited. As noted by Singpurwalla (2006, p. 17), the concept of frequentist probabilities "is applicable to only those situations for which we can conceive of a repeatable experiment". This excludes many situations and events. Think of the rise of the sea level over the next twenty years, the guilt or innocence of an accused individual, or the occurrence or not of a disease in a specific person with a specific history.

What does it mean that the situations considered are 'similar'? The conditions under which the repeatable experiments are to be performed cannot be identical, because in that case we would get exactly the same outcome and the ratio n_A/n would be either 1 or 0. What type of variation from one experiment to another is allowed? This is often difficult to specify and makes it hard to extend frequentist probabilities to include real-life situations. Think, for example, of the frequentist probability that a person V will get a specific disease. What should then be the population of similar persons? If we include all men/women of his/her age group we get a large population, but many of the people in this population may not be very 'similar' to person V. We may reduce the population to increase the similarity, but not too much because that would make the population very small and hence inadequate for defining the frequentist probability. The issue raised here can be seen as a special case of the 'reference class' problem (Hajek 2007).

We face this type of dilemma in many types of modelling. A balance has to be struck between different concerns. We cannot be too person-specific if we are to be able to estimate the parameters of the model accurately.

Let us consider the operation of an offshore oil and gas installation. As an example, we might contemplate introducing a frequentist probability that an accident with more than one hundred fatalities occurs in the next year. What the infinite population of similar situations (platforms) is to reflect to define this probability is not at all clear, as discussed in Section 2.2.1; see also Aven (2012a, p. 37). Does 'similar' mean the same type of constructions and equipment, the same operational procedures, the same type of personnel positions, the same type of training programmes, the same organisational philosophy, the same influence of exogenous factors, etc.? "Yes" would be the answer as long as we speak about similarities on a macro level. However, as discussed above, something must be different, because otherwise we would get exactly the same output result for each situation (platform) – either the occurrence of an accident with at least one hundred fatalities or no such event. There must be some variation at a micro level to produce the variation on the macro level. So we should allow for variations in equipment quality, human behaviour and so on. However, the question is to what extent and how we should allow for such variation. For example, in human behaviour, do we specify the safety culture or the standard of the private lives of the personnel, or are these factors to be regarded as those creating the variations from one situation to another, i.e. the stochastic (aleatory) uncertainty, using common terminology within risk analysis (see Section 3.1)? We see that we will have a hard time specifying what the framework conditions (i.e. the conditions that are fixed) should be for the situations in the population, and what could be variable.

As seen from this example, it is not obvious how to obtain a proper definition of the population, and hence define and understand the meaning of the frequentist probability. Clearly such a situation could also have serious implications for the estimation of the probabilities: trying to estimate a concept that is vaguely defined and understood would easily lead to arbitrariness in the assessment.

If frequentist probabilities had been taken as the basis for the assessment in the offshore case described in Section 1.3, the analysts would have reported estimates of the underlying frequentist probabilities. The managers should have been informed about this and also about the issue of uncertainties in these estimates relative to the underlying presumed 'true' frequentist probabilities. This was not done in the actual case, and it would have been a serious weakness of the assessment if the frequentist probability framework had been the basis for the analysis: such a framework requires uncertainty analysis of the estimates if it is to be complete. The topic is thoroughly discussed in Aven (2011a); see also Section 3.3.4 when referring to the probability of frequency approach. It relates to traditional statistical methods as well as to the use of subjective probabilities to assess the uncertainties about the 'true' value of the underlying frequentist probabilities.

A.1.3 Logical probabilities

This type of probability was first proposed by Keynes (1921) and later taken up by Carnap (1922, 1929). The idea is that probability expresses an objective logical relation between propositions, a kind of 'partial entailment'. There is a number in the interval [0,1], denoted P(h|e), which measures the objective degree of logical support that evidence e gives to the hypothesis h (Franklin 2001). As stated by Franklin (2001), this view on probability has an intuitive initial attractiveness, in representing a level of agreement found when scientists, juries, actuaries and so on evaluate hypotheses in the light of evidence. However, the notion of partial entailment has never received a satisfactory interpretation as argued by, for example, Cowell *et al.* (1999) and Cooke (2004). Using logical probabilities, it is not clear how we should interpret a number (say) 0.2 compared to 0.3. Hence this conceptualisation of probabilities is not suitable for applications.

A.2 'Subjective' probabilities

The concept of subjective probability and two of the most common interpretations were introduced in Section 2.2.1. Let us study these and related definitions in a little more detail. First, betting and related types of interpretations are addressed, followed by the interpretation based on an uncertainty standard such as an urn.

A.2.1 Betting and related types of interpretations

The subjective theory of probability was proposed independently and at about the same time by Bruno de Finetti in Italy in *Fondamenti Logici del Ragionamento Probabilistico* (1930) and by Frank Ramsey in Cambridge in *The Foundations of Mathematics* (1931); see Gillies (2000). In line with de Finetti, the probability of the event A, P(A), equals the amount of money that the assigner would be willing to put on the table if he/she were to receive a single unit of payment in the case that the event A were to occur, and nothing otherwise. The opposite must also hold, i.e. the assessor must be willing to pay the amount $1 - P(A)$ if he/she were to receive a single unit of payment in the case that A were not to occur, and nothing otherwise. An alternative formulation expressing the same idea is: the probability of an event is the price at which the person assigning the probability is neutral between buying and selling a ticket that is worth one unit of payment if the event occurs, and worthless if not (Singpurwalla 2006).

Let us see what this means in a concrete case. Consider the event A, defined as the occurrence of a nuclear accident (properly specified). Suppose you assign the subjective probability P(A) = 0.005. You are expressing your indifference between receiving (paying) €0.005 or taking a gamble in which you receive (pay) €1 if A occurs and €0 if A does not occur. If the unit of money is €1000, the interpretation would be that you are indifferent between receiving (paying) €5 or taking a gamble where you receive (pay) €1000 if A occurs and €0 if A does not occur. In practice, you may assign the value by an iterative procedure such as this (using the unit €1000):

To take the gamble, the question is: "are you willing to pay more or less than €100?" You answer "less", and the second question is: "are you willing to pay more or less than €10?" You answer "less", and the next question is: "are you willing to pay more or less than €1?" You answer "more", and the procedure continues until you are indifferent. Then your probability is specified.

Think about this example. Would you take this type of gamble? Given that the nuclear accident occurs, you will get €1000. However, as noted by Lindley (2006) (and see also Cooke [1986]), receiving the payment would be trivial if the accident were to occur (the assessor might not be alive to receive it). The problem is that there is a link between the probability assignment and value judgements concerning the price of the gamble, the money. How important is the €1000 for you? This value judgement has nothing to do with the uncertainties per se, or your degree of belief in the event A occurring. We see that the interpretation is based on a mixture of uncertainty assessments and value judgements.

This fact makes this way of looking at subjective probabilities problematic to use in practice. If you are a decision-maker and would like to be informed by an expert expressing his or her subjective probability about an event A, you would not be happy about the probability being influenced by this expert's attitude to the gambling situation and the money involved.

Let us consider the offshore installation case again. Assume that the analysts used subjective probabilities and a betting type of interpretation, and say that a figure of 1/10000 was assigned for a probability of an accident with at least one hundred fatalities during a specific period of time. Following this interpretation, the meaning of this assignment is that the analysts are indifferent between receiving 1/10000 units of money and playing a gamble where the reward is 1 if the event occurs and 0 otherwise. However, if such an interpretation had been used, the management would clearly have protested, because the risk analysts have entered their domain as managers and decision-makers by incorporating in the probability assignment their own judgements of the value of money. The purpose of the risk assessment is to inform the management about the safety on the platform, but the suggested approach makes it impossible for the management to evaluate the results in a meaningful way, because the numbers produced by the analysts are disturbed by the analysts' value judgements which are irrelevant to the decision-making. The approach conflicts with the basic principle of a separation between analysis and value judgements, which is fundamental in the risk analysis and safety field when management are not making the analysis judgements themselves.

Given these problems, one may question why this type of interpretation has the appeal that it seems to have, looking at the probability literature. One reason is evident: subjective probability theory is strongly associated with the pioneers, Bruno de Finetti, Frank Ramsey and Leonard Savage. Although presenting different frameworks for understanding the concept of subjective probability, these authors share a common feature: probability assignments are linked to value judgement about money and other attributes, as will be clear from the following brief summary of Ramsey's and Savage's ideas.

Ramsey argues that people's beliefs can be measured by the use of a combined preference-utility method. Following Cooke (1986), the method is based on establishing an equivalence in preference between a lottery on an event A with payoffs in money, and an amount of money. Putting the lottery in curly brackets, suppose we observe that {€x if A and €y if not A} is equivalent in preferences to a fixed amount, €z. Then the probability of A is given by:

$$p(A) = [U(z) - U(y)]/[U(x) - U(y)]$$

where U denotes a utility function on money. This type of probability would suffer from the same type of problems as in the de Finetti type of interpretation considered above; see Cooke (1986). We need not go into the details, as we immediately see that we again have a mixture of uncertainty assessments and value judgements. Cooke refers to this method for defining a subjective probability as 'vulgar Ramsey', and he also discusses a 'non-vulgar' version, but, for the purpose of the present review, it suffices to address the 'vulgar Ramsey'. For the 'non-vulgar' version, the probabilities are also linked to utilities.

There are a number of other definitions of subjective probabilities which can be grouped in the same categories as those above, because they mix uncertainty assessments and value judgements. Savage's (1956) theory deserves to be mentioned in this context. Savage has worked out a comprehensive system for understanding probability on the basis of *preferences* between acts; see Bedford and Cooke (2001). By observing a number of choices in preferences (for example, "€100 if it rains tomorrow, otherwise €0" and "€100 if the Dow-Jones goes down tomorrow, otherwise €0"), the assigner's subjective probabilities can be determined. Making some general assumptions about the assigner's utility function, for example that it is linear, the probabilities can be easily deduced, as shown in Bedford and Cooke (2001, p. 31).

These three pioneers, but also others, for example Borel (1950) and Jeffrey (2004), have stimulated a lot of work on subjective probabilities linked to betting, utilities and value judgements in general. The literature includes a number of papers based on these ways of looking at subjective probabilities. However, there are relatively few examples of scholars challenging these perspectives. Among advocates of subjective probabilities, few seem critical of these perspectives. Cooke (1986) provides a critical analysis of the definitions of de Finetti and Ramsey, but does not seem to have problems in adopting Savage's preference approach (Bedford and Cooke 2001). Looking at the larger population of researchers in probability, many other names could be mentioned, for example Good (1951). He writes (p. 110):

> It is possible, and quite usual, to discuss probability with little reference to utilities. If utilities are introduced from the start, the axioms are more complicated and it is debatable whether they are more "convincing". The plan which appeals to me is to develop the theory of probability without much reference to utilities, and then to adjoin the principles of rational behavior in order to obtain a theory of rational behavior.

Good (1951) relates probability to degrees of belief but does not provide any interpretation of this concept (see pp. 108–109). It is common to refer to Good's perspective as a logical Bayesian one, as he sees subjective probability as epistemic in the sense that it results from an evaluation based on the available evidence and does not depend on the assessor's state of mind. We may thus place Good in the category of logical probabilists.

As mentioned above, however, an alternative interpretation is provided by Dennis Lindley (see e.g. Lindley 1985, 2000, 2006), who strongly argues for a separation between probability and utility (see Sections 2.2.1 and 5.1.1). It seems that many authors have not properly understood Lindley's interpretation – this may be another explanation of why the betting interpretation dominates the literature.

Finally in this section, a remark on the term 'chance' used in the Bayesian literature. A chance is the limit of a relative frequency in an exchangeable, infinite Bernoulli series (Lindley 2006). The relationship between a subjective probability and chance is given by de Finetti's so-called representation theorem; see e.g. Bernardo and Smith (1994, p. 172). Roughly speaking, this theorem states that if the exchangeable Bernoulli series can be justified, the analyst can act as if frequency probabilities exist. Conditional on the chance, the Bernoulli random quantities are independent. In this case we are led to a framework (the Bayesian framework) where probability is subjective and chances are introduced to represent variation in populations (also referred to as aleatory uncertainties).

A.2.2 The reference to an uncertainty standard

There are fairly few examples of researchers and probabilists who refer to Lindley's standard uncertainty interpretation of a subjective probability (one key reference is Kaplan and Garrick [1981]). This is surprising; an elegant, simple and easily understandable foundation and theory for subjective probability is available but is not applied. Let us look a little more closely at this type of interpretation.

According to this definition, a subjective probability is understood in relation to an uncertainty standard, typically an urn: if a person assigns a probability of 0.1 (say) for an event A, he or she compares his/her uncertainty (degree of belief) of A occurring with the likelihood of drawing a specific ball from an urn containing ten balls. The uncertainty (degree of belief) is the same. From this standard we can deduce a set of rules, for example the additive rule (Lindley 2000, p. 296): "Suppose that, of the N balls in the urn, R are red, B are blue, and the remainder are white. Then the uncertainty that is associated with the withdrawal of a colored ball is $(R+B)/N = R/N + B/N$, the sum of the probabilities/uncertainties associated with red, and with blue, balls. The same result we will obtain for any two exclusive events whose probabilities/uncertainties are respectively R/N and B/N and we have an addition rule for your probabilities/uncertainties of exclusive events." Similarly, we can establish other common rules for probability, for example the multiplicative rule; see Lindley (2000, p. 298).

These rules are commonly referred to as axioms in text books on probability, but they are not axioms here, since they are deductions from the more basic assumptions linked to the uncertainty standard; see Lindley (2000, p. 299).

For applied probabilists, whether the probability rules are deduced or considered to be axioms may not be so important. The main point is that these rules apply, and hence Lindley's uncertainty standard provides an easily understandable way of defining and interpreting subjective probability, based on a separation between uncertainty/probability and utility/values. The rules of probability reduce to the rules governing proportions, which are easy to communicate.

The Bayesian framework introduced in the previous section, with subjective probabilities and chances, can of course also be used when the uncertainty standard is adopted.

Aven (2013g) refers to a reviewer who indicated that the 'uncertainty standard' suffers from all the problems that the author attributes to the 'classical interpretation' and the 'frequentist probabilities' – the author's presentation is said to perform a circle and ends up with an interpretation of probability that is based on the concept of 'objective probability'. The reviewer requested a better explanation of the difference between the subjective probability interpreted using the uncertainty standard (Lindley 2000) and the classical definition.

As a response to these comments, let us remind ourselves that a subjective probability (regardless of interpretation) is not an objective probability, since it expresses the assigner's judgement (a degree of belief) related to the occurrence of the event of interest. There is no circle in the presentation as indicated by this reviewer. The fact that a classical probability setting (the urn model) is used as a reference system to explain what subjective probabilities mean, does not make the subjective probability in any way objective. It still expresses the subjective judgements of the assigner. As an example, consider the event A = "the sea level will increase by one metre during the coming thirty years", and say that an analyst assigns a subjective probability of event A equal to 0.1. Following the Lindley interpretation, the meaning of this statement is that the analyst's uncertainty/degree of belief that the event A will occur is the same as for the event B = "drawing one particular ball out of an urn that consists of ten balls" (under standard experimental conditions). Although there is a comparison with classical probability, the subjective probability of A is not objective. The classical probability introduced here is not a classical probability of the event A, but a constructed event B introduced to explain what the probability of A means.

In contrast, to adopt a classical probability for the event A, we have to define a finite number of outcomes which are equally likely to occur, and the probability of A is equal to the ratio between the number of outcomes resulting in A and the total number of outcomes. However, defining such outcomes is obviously problematic in this case – we cannot meaningfully construct similar thirty-year periods, and we therefore have to conclude that classical probability has no proper interpretation for this event. We also have to reach the same conclusion for frequentist probabilities; an infinite population of similar thirty-year periods cannot be meaningfully defined. Nonetheless, the analyst may assign subjective probabilities, and to explain what these mean, the urn model is a suitable tool.

There is no reference here to an underlying correct or true value (the same is the case for the betting type of interpretations). The assigned probabilities express the judgements of the assessors. However, the probabilities are based on some background knowledge (covering assumptions, models and data), and this may be more or less strong. This raises the question about the information value given by the subjective probabilities produced. The probability numbers must be viewed together with the background knowledge on which the probabilities are based. This issue has motivated the research on alternative approaches, including imprecise probabilities, as discussed in the following section.

A.3 Alternative representations

A.3.1 Imprecise (interval) probabilities

We all deal with imprecise (interval) probabilities in practice. If the assessor assigns a probability P(A) = 0.3, one may interpret this probability as having an imprecision interval [0.26, 0.34] (since a number in this interval is equal to 0.3 when displayed to one significant figure). More formally, we may define imprecise probabilities (or interval probabilities) for event A in this way:

Uncertainty is represented by a lower probability $\underline{P}(A)$ and an upper probability $\overline{P}(A)$, giving rise to a probability interval $[\underline{P}(A), \overline{P}(A)]$, where $0 \leq \underline{P}(A) \leq \overline{P}(A) \leq 1$. The difference

$$\Delta P(A) = \overline{P}(A) - \underline{P}(A) \tag{A.1}$$

is called the imprecision in the representation of the event A.

Following the uncertainty standard interpretation, this means that the assigner states that his/her assigned degree of belief is greater than the urn chance of 0.26 (the degree of belief about drawing one red ball out of an urn containing 100 balls where 26 are red) and less than the urn chance of 0.34. The analyst is not willing to make any further judgements. It is also possible to give an interpretation in line with de Finetti's interpretation: the lower probability is interpreted as the maximum price at which one would be willing to place a bet which pays 1 if A occurs and 0 if not, and the upper probability as the minimum price for which one would be willing to place the same bet (see Walley 1991).

Following the arguments in the previous section, only the former interpretation can be considered adequate for applied contexts. The latter interpretation is rejected because it mixes uncertainty assessments and value judgements.

Lindley (2000) argues that the interval approach leads to a more complicated system, and the complication seems unnecessary. He has not yet met a situation in which the probability approach appears to be inadequate and where the inadequacy can be fixed by employing upper and lower values. The simpler is to be preferred over the more complicated, he concludes. Furthermore, he argues that the use of interval probabilities confuses the concept of interpretation with the practice of measurement procedures (Lindley 2006). To use the words of Bernardo and Smith (1994, p. 32), the idea of a formal incorporation of imprecision into the axiom system represents "an unnecessary confusion of the *prescriptive* and the *descriptive*":

> We formulate the theory on the prescriptive assumption that we aspire to exact measurement (...), whilst acknowledging that, in practice, we have to make do with the best level of precision currently available (or devote some resources to improving our measuring instruments!).
>
> (Bernardo and Smith 1994, p. 32)

However, other researchers and analysts have a more positive view on the need for such intervals; for example, see discussions in Aven *et al.* (2014a), Walley (1991), Ferson and Ginzburg (1996) and Dubois (2010). See also Section 3.3.1.

A.3.2 Probability bound analysis

Ferson and Ginzburg (1996) present a combined probability analysis – interval analysis – referred to as a probability bound analysis. The context is a risk assessment where the purpose is to express uncertainties about some parameters θ_i of a model. For the parameters where the aleatory uncertainties cannot be accurately estimated, probability intervals are used. In this way the uncertainty analysis is carried out in the traditional probabilistic way for some parameters, and intervals are used for others. More specifically this means that:

(1) For parameters θ_i where the aleatory uncertainties cannot be accurately estimated, use interval analysis where $a_i \leq \theta_i \leq b_i$ for constants a_i and b_i.
(2) For parameters θ_i where the aleatory uncertainties can be accurately assessed, use frequentist probabilities to describe the distribution over θ_i.
(3) Combine 1 and 2 to generate a probability distribution over θ which is a function of the parameters θ_i, for the different interval limits. For example, assume that for i = 1, an interval is used with bounds a_1 and b_1, whereas for i = 2, a probabilistic analysis is used. Then we obtain a probability distribution over $\theta = \theta_1 \theta_2$ (say) when $\theta_1 = a_1$ and a probability distribution over θ when $\theta_1 = b_1$.

Following this approach, subjective probabilities are not used. Bounds replace the epistemic-based probabilities.

A.3.3 Possibility theory

In accordance with possibility theory, uncertainty is represented by using a possibility function r(x). For each x in a set Ω, r(x) expresses the degree of possibility of x. When r(x) = 0 for some x, the outcome x is considered an impossible situation. When r(x) = 1 for some x, the outcome x is possible, i.e. is unsurprising, normal, usual (Dubois 2006). This is a much weaker statement than when the probability is 1.

The possibility function r gives rise to probability bounds, upper and lower probabilities, referred to as the necessity and possibility measures (Nec, Pos). They are defined as follows.

The possibility (plausibility) of an event A, Pos(A), is defined by:

$$Pos(A) = \sup_{\{x \in A\}} r(x) \tag{A.2}$$

and the necessity measure Nec(A) is defined by:

$$Nec(A) = 1 - Pos(A^c)$$

where A^c is the complement of A.

Let P(r) be a family of probability distributions such that for all events A,

$$Nec(A) \leq P(A) \leq Pos(A)$$

Then:

$$Nec(A) = \inf P(A) \text{ and } Pos(A) = \sup P(A) \tag{A.3}$$

where inf and sup are with respect to all probability measures in P. Hence the necessity measure is interpreted as a lower level for the probability, and the possibility measure is interpreted as an upper limit. Using subjective probabilities, the bounds reflect that the analyst is not able or willing to precisely assign his/her probability. He or she can only describe a subset of P which contains his/her probability (Dubois and Prade 1989).

A typical example of possibilistic representation is the following (Anoop and Rao 2008, Baraldi and Zio 2008). We consider an uncertain parameter x. Based on its definition, we know that the parameter can take values in the range [1,3] and the most likely value is 2. To represent this information, a triangular possibility distribution on the interval [1, 3] is used, with maximum value at 2; see Figure A.1.

From the possibility function we define α cut sets F_α = {x: r (x) $\geq \alpha$}, for 0 $\leq \alpha \leq 1$. For example, $F_{0.5}$ = [1.5, 2.5] is the set of x values for which the possibility function is greater than or equal to 0.5. From the triangular possibility distribution in Figure A.1, we can conclude that if A expresses that the parameter lies in the interval [1.5, 2.5], then 0.5 $\leq$ P(A) $\leq$ 1.

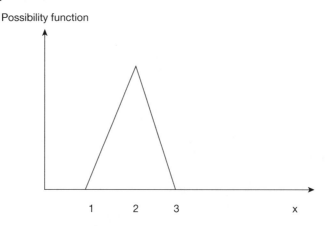

Figure A.1 Possibility function for a parameter on the interval [1, 3], with maximum value at 2 (based on Aven and Zio 2011)

From Equation (A.3) we can deduce the associate cumulative necessity/possibility measures Nec(-∞, x] and Pos(-∞, x], as shown in Figure A.2. These measures are interpreted as the lower and upper limiting cumulative probability distributions for the uncertain parameter x. Hence the bounds for the interval [1, 2] are $0 \le P(A) \le 1$.

These bounds can be interpreted as for the interval probabilities: the interval bounds are those obtained by the analyst when he/she is not able or willing to precisely assign his/her probability – the interval is the best he/she can do given the information available.

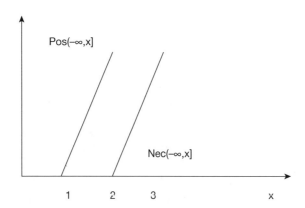

Figure A.2 Bounds for the probability measures based on the possibility function in Figure A.1 (based on Aven and Zio 2011)

A.3.4 Evidence theory (Dempster-Shafer theory, the theory of belief functions)

Random sets in the two forms proposed by Dempster (1967) and Shafer (1976) are based on the specification of *beliefs* and *plausibilities*, for each subset of outcomes (events) in the sample space under consideration. This allows the theory to take into account the weight of evidence. Possibility theory can be considered a special case of this theory.

Consider an event A and its complement A^c. These are mutually exclusive and exhaustive events, and in probability theory their respective probabilities are required to sum to one. Thus, if the event A is assigned the probability p, then A^c must be assigned the probability $1 - p$. On the contrary, in evidence theory degrees of belief are assigned based on the strength of the supporting evidence: the belief value must represent the degree to which evidence is judged to support a given proposition, and the degree of belief is explained by Shafer (1976) as the commitment of a certain portion of someone's belief. If there is little evidence both in favour of and against the event A, then the beliefs in both its occurrence and its non-occurrence should be assigned low values. In the extreme case of no evidence at all, both beliefs should be set to zero. Letting $Bel(A)$ denote the degree of belief that A will occur and $Bel(A^c)$ the degree of belief that A will not occur, the requirement is only that

$$Bel(A) + Bel(A^c) \leq 1$$

Thus, the specification of the belief function is capable of incorporating a lack of confidence in the occurrence of the event A, quantitatively manifested in the sum of the beliefs of the occurrence ($Bel(A)$) and non-occurrence ($Bel(A^c)$) being less than one. The difference $1 - [Bel(A) + Bel(A^c)]$ is called *ignorance*. When the ignorance is 0, the available evidence justifies a probabilistic description of the uncertainty.

According to Shafer (1976), an adequate summary of the impact of evidence must include at least two items of information: the support of the evidence in favour and the support of the evidence against. The plausibility of the event A, $Pl(A)$, is then introduced as the extent to which evidence does not support A^c and the relation between plausibility and belief is

$$Pl(A) = 1 - Bel(A^c)$$

A fundamental property of the plausibility function is that:

$$Pl(A) + Pl(A^c) \geq 1$$

Thus, the specification of the plausibility function reflects the evidence in support of the occurrence or not of the event A, as quantified by the sum of the plausibilities of the occurrence ($Pl(A)$) and non-occurrence ($Pl(A^c)$) being greater than or equal to one.

The theory is based on the idea of obtaining degrees of belief for one question from subjective probabilities for related questions (Shafer 1990). To illustrate, suppose that a diagnostic model is available to indicate with reliability (i.e. the probability of providing the correct result) of 0.9 when a given system has failed. Considering a case in which the model does indeed indicate that the system has failed, this fact justifies a 0.9 degree of belief in such an event but only a 0 degree of belief (not 0.1) on the event that the system has not failed. This latter belief does not mean that it is certain that the system has failed, as a zero probability would; it merely means that the model indication provides no evidence to support the fact that the system has not failed. The pair of values {0.9, 0} constitutes a belief function on the propositions "the system has failed" and "the system has not failed".

From the above simple example, one can appreciate how the degrees of belief for one question (has the system failed?) are obtained from probabilities related to another question (is the diagnostic model correct?).

Denoting the event that the system has failed as A and the diagnostic indication of the system state as m, the conditional probability $P(m|A)$, i.e. the model reliability, is used as the degree of belief that the system has failed. This is unlike the standard Bayesian analysis, where the focus would be on the conditional probability of the failure event given the state diagnosis by the model, $P(A|m)$, which is obtained by updating the prior probability of A, $P(A)$, using Bayes' rule.

As for the interpretation of the measures introduced in evidence theory, Shafer (1990) uses several metaphors for assigning (and hence interpreting) belief functions. The simplest says that the assessor judges that the strength of the evidence indicating that the event A is true, $Bel(A)$, is comparable with the strength of the evidence provided by a witness who has a $Bel(A)$ 100% chance of being reliable. Thus, we have

$Bel(A) = P$(The witness claiming that A is true is reliable)

The metaphor is to be interpreted as the diagnostic model analysed above, with witness reliability playing the role of model reliability.

A.4 Conclusions

To apply a probabilistic analysis in practice, an interpretation is required. You have to explain what the probability means. For example, is your probability to be understood as a judgement made by the analyst team based on their background knowledge, or is the probability trying to represent the data and knowledge available in a more 'objective' way? The interpretation could affect the decision-making process significantly.

Many interpretations of probability exist, but only a few are meaningful in practice. Above, it is argued that there is only one way of looking at probability that is universally applicable, provided we require a separation between uncertainty assessments per se and value judgements: the subjective probabilities interpreted with reference to an uncertainty standard (for example an urn). The value-based interpretations of a subjective probability must be rejected because they are based on an unfortunate mix of uncertainty assessment and value judgements. It should be possible to define a probability without having to specify one's utility function.

Many applied probabilists find the term 'subjective probability' difficult to use in practice – it gives the impression that the probability and the associated assessment are non-scientific and arbitrary, and so it is often replaced by terms such as 'judgemental probability' and 'knowledge-based probability' (Singpurwalla 2006, North 2010, Aven 2010b).

In a communication process, frequencies can also be used to explain the meaning of a subjective probability, provided it makes sense to talk about situations similar to the one studied. If the subjective probability assignment is 0.3, the analyst predicts that the event considered will occur three times out of ten. Note that this way of speaking does not imply that a frequentist probability needs to be defined – the frequency referred to is simply introduced to give a feel for what the assignment of 0.3 expresses.

Frequentist probabilities (chances) $P_f(A)$ may be constructed in cases of repeatability. These probabilities are to be seen as parameters of a probability model. When a frequentist probability can be justified, for convenience we may also accept the concept of a propensity, that the probability exists per se; the probability is just a propensity of a repeatable experimental set-up which produces outcomes with limiting relative frequency $P_f(A)$.

Classical probabilities only exist in some special cases of gambling and sampling. Consequently, their applicability is very limited. Logical probabilities cannot be justified. Imprecise probabilities can be meaningfully defined using an uncertainty standard, but their applicability may be questioned in some cases. The approach may complicate the assessments, and if the aim of the analysis is to report the analyst's judgement, exact probabilities should be assigned. However, if we seek to obtain a more 'inter-subjective' knowledge description of the unknown quantities studied, the use of imprecise probabilities (for example, founded on possibility theory or evidence theory) could be an appropriate tool, as discussed above.

The overall conclusion is that in a risk analysis and safety setting, a probability should in general be interpreted as a subjective probability with reference to an uncertainty standard such as an urn. In addition we may use frequentist probabilities (chances) and classical probabilities when these can be justified. Imprecise probabilities can also be used in cases where it is difficult to assign specific numbers, interpreted in line with the reference to an uncertainty standard.

Appendix B
A summary of risk definitions

It is impossible to present and discuss all the definitions of the risk concept suggested and used in the scientific risk fields. Several definitions can be found in, for example, Wood (1964), Crowe and Horn (1967), Aven and Renn (2009) and Aven *et al.* (2011), but there are many more. However, by structuring and classifying the risk definitions according to some specific features as shown below, it is possible to cover a large number of definitions.

The references in parentheses specify the sources of the relevant definitions except for 1b, 1d, 2b and 6d, where the references simply provide examples of work in which the relevant definitions are used or referred to.

1   Risk = Expected value (loss) (R = E)
    a   The risk of losing any sum is the reverse of expectation, and the true measure of it is the product of the sum adventured multiplied by the probability of the loss (De Moivre 1711).
    b   Risk equals the expected loss (Verma and Verter 2007, Willis 2007).
    c   Risk equals the product of the probability and utility of some future event (Adams 1995).
    d   Risk equals the expected disutility (Campbell 2005).

2   Risk = Probability of an (undesirable) event (R = P)
    a   Risk is the chance of damage or loss (Haynes 1895).
    b   Risk equals the probability of an undesirable event (Campbell 2005).
    c   Risk means the likelihood of a specific effect originating from a certain hazard occurring within a specified period or in specified circumstances (Kirchsteiger 2002).

3   Risk = Objective uncertainty (R = OU)
    a   Risk is the objective correlative of the subjective uncertainty; with uncertainty considered as embodied in the course of events in the external world (Willett 1901).
    b   Risk is measurable uncertainty, i.e. uncertainty where the distribution of the outcome in a group of instances is known (either

through calculation a priori or from statistics of past experience) (Knight 1921).

4   Risk = Uncertainty (R = U) (Angell 1959, Mowbray and Blanchard 1961)
    a   In regard to cost, loss or damage (Hardy 1923).
    b   About a loss (Mehr and Cammack 1953).
    c   About the happening of an unfavourable contingency (Magee 1961).
    d   Of outcome, of actions and events (Cabinet Office 2002).

5   Risk = Potential/possibility of a loss (R = PO)
    a   Risk is the possibility of an unfortunate occurrence (Riegel and Miller 1966).
    b   Risk is the possibility of an unfavourable deviation from expectations (Atheam 1969).
    c   Risk is the potential for the realisation of unwanted, negative consequences of an event (Rowe 1977).

6   Risk = Probability and scenarios/consequences/severity of consequences (R = (P,C))
    a   Risk is a combination of hazards measured by probability; a state of the world rather than a state of mind (Pfeffer 1956).
    b   Risk is a measure of the probability and severity of adverse effects (Lowrance 1976).
    c   Risk is equal to the triplet (s_i, p_i, c_i), where s_i is the ith scenario, p_i is the probability of that scenario, and c_i is the consequence of the ith scenario, i = 1, 2, ...N; i.e., risk captures: What can happen? How likely is that to happen? If it does happen, what are the consequences? (Kaplan and Garrick 1981).
    d   Risk is the combination of probability and extent of consequences (Ale 2002).

7   Risk = Event or consequence (R = C)
    a   Risk is a situation or event where something of human value (including humans themselves) is at stake and where the outcome is uncertain (Rosa 1998, 2003).
    b   Risk is an uncertain consequence of an event or an activity with respect to something that humans value (IRGC 2005).

8   Risk = Consequences/damage/severity of these + uncertainty (R = (C,U))
    a   Risk = uncertainty + damage (Kaplan and Garrick 1981).
    b   Risk is equal to the two-dimensional combination of events/consequences (of an activity) and associated uncertainties (Aven 2007, Aven 2010b).

    c   Risk is uncertainty about and severity of the consequences (or outcomes) of an activity with respect to something that humans value (Aven and Renn 2009).
    d   Risk is the deviations from a reference level (ideal states, planned values, expected values, objectives) and associated uncertainties (Aven and Aven 2011).

9   Risk is the effect of uncertainty on objectives (ISO 2009a, b) (R = ISO).

Appendix C
Terminology

This appendix summarises the risk analysis and management terminology used in the book. Unless otherwise stated, the terminology is in line with the international guidelines (ISO 2009a). For some comments on these guidelines, see Aven (2011f).

- *aleatory (stochastic) uncertainty*: variation of quantities in a population.
  This definition is not given in the ISO guidelines.
- *epistemic uncertainty about something*: not knowing about something, where 'something' refers to the true value of a quantity or the true future consequences of an activity.
  This definition is not given in the ISO guidelines.
- *event*: occurrence or change in a particular set of circumstances.
- *frequency*: number of events per unit of time or another reference. Often frequency is also used for the expected number of events per unit of time.
- *hazard*: A risk source or an associated event where the consequences relate to harm.
- *managerial review and judgement*: process of summarising, interpreting and deliberating over the results of risk assessments and other assessments, as well as of other relevant issues (not covered by the assessments), in order to make a decision.
  This definition is not given in the ISO guidelines.
- *probability*: either a knowledge-based (subjective) measure of uncertainty about an event conditional on some background knowledge, or a frequentist probability (chance). If a knowledge-based probability is equal to 0.10, it means that the uncertainty (degree of belief) is the same as randomly drawing a specific ball out of an urn containing ten balls. A frequentist probability (chance) is the fraction of events A occurring when the situation under consideration can be repeated over and over again infinitely.
  This definition is not given in the ISO guidelines.

- *risk*: the book refers to different definitions. The most general and the one recommended states that, in relation to an activity, risk is the two-dimensional combination of:

  i   consequences C of the activity (with respect to something that humans value);
  ii   associated uncertainties about C (C being unknown).

  This definition is referred to as the (C,U) definition.
  This definition is not given in the ISO guidelines.
- *risk acceptance*: a decision to accept risk.
  This definition represents an adjustment of the definition used by the ISO guidelines: informed decision to take a particular risk.
- *risk acceptance criterion*: a reference by which risk is assessed to be acceptable or unacceptable.
  This definition is not included in the ISO guidelines. A risk acceptance criterion is an example of a risk criterion.
- *risk analysis*: systematic use of data, information and knowledge to identify risk sources, causes and consequences of these sources, and to describe risk.
  The ISO guidelines do not include source identification as a part of risk analysis. The guidelines state that a risk analysis is the process undertaken in order to comprehend the nature of risk and to determine the level of risk.
  Risk analysis is also often understood in a broader way (Thompson *et al.* 2005): risk analysis is defined to include risk assessment, risk characterisation, risk communication, risk management and policy relating to risk, in the context of risks of concern to individuals, to public and private sector organisations, and to society at a local, regional, national, or global level. This definition is discussed in Aven (2012e).
- *risk appetite*: amount and type of risk an organisation is willing to take on risky activities in pursuit of values.
- *risk assessment*: the overall process of risk analysis and risk evaluation.
- *risk communication*: exchange or sharing of risk-related data, information and knowledge between stakeholders.
  This definition is not given in the ISO guidelines.
- *risk criteria*: terms of reference against which the significance of the risk is evaluated.
- *risk description*: a qualitative and/or quantitative picture of the risk; i.e. a structured statement of risk usually containing the elements: risk sources, causes, events, consequences and uncertainty representations/ measurements.

Using the set-up discussed in Sections 2.3 and 2.4, risk is described by (C',Q,K), where C' are the defined consequences of the activity considered, Q the measure of uncertainty used, and K the background knowledge that C' and Q are based on.

This definition represents an adjustment of the one given in the ISO guidelines.

- *risk evaluation*: process of comparing the result of risk analysis against risk criteria to determine the significance of the risk.

  This definition represents an adjustment of the definition used by the ISO guidelines.

  See also managerial review and judgement.

- *risk level*: assessed magnitude of the risk.

  This definition is not given in the ISO guidelines.

- *risk management*: coordinated activities to direct and control an organisation with respect to risk.

- *risk perception*: stakeholder's subjective judgement or appraisal of risk.

  This definition represents an adjustment of the definition used by the ISO guidelines.

- *risk quantification*: process used to assign values to the probabilities and risk indices used.

  This definition is not given in the ISO guidelines.

- *risk retention*: acceptance of the potential benefit of gain, or burden of loss, from the risk.

- *risk source*: element which alone or in combination has the intrinsic potential to give rise to an event with a consequence.

  This definition represents an adjustment of the definition used by the ISO guidelines.

- *risk tolerability level*: level of risk which an organisation will tolerate.

  This definition is not given in the ISO guidelines.

- *risk treatment*: process to modify risk.

- *stakeholder*: person or organisation that can affect, be affected by, or perceive themselves to be affected by a decision or activity.

- *threat*: a risk source or an associated event.

- *uncertainty about something*: not knowing about something, where 'something' refers to the true value of a quantity or the true future consequences of an activity.

  This definition is not given in the ISO guidelines.

- *uncertainty description*: a measure of the uncertainty and associated background knowledge, i.e. (Q,K) using the notation defined above for risk description.

  This definition is not given in the ISO guidelines.

- *vulnerability*: in line with the (C,U) risk definition, the vulnerability given an event A or a risk source RS, is the two-dimensional combination of:

i   consequences C of the activity (with respect to something that humans value);

ii   associated uncertainties about C (C being unknown)

given the occurrence of A or RC. We write (C,U | A/RS). Vulnerability is generally described as (C',Q,K | A/RS).

This definition is not given in the ISO guidelines.

Appendix D
Calculations of the probability of a black swan (Lindley)

(a) The first part of this appendix covers the black swan example presented by Lindley (2008) and shows how he arrived at his results.

Let p be the fraction of successes (white swans) in the infinite series of trials (the total population of swans), and let X_n be the number of successes (white swans) in n trials (swans). Furthermore, let Y_m be the number of successes (white swans) in m new trials (swans). We will compute the probability of m successes in the new trials given only successes in the first n trials, i.e. $P(Y_m = m| X_n = n)$. By conditioning on the true value of p, we find that

$$P(Y_m = m| X_n = n) = \int_{[0,1]} P(Y_m = m| X_n = n, p) \, dH(p| X_n = n)$$
$$= \int_{[0,1]} P(Y_m = m| p) \, dH(p| X_n = n) = \int_{[0,1]} p^m \, dH(p| X_n = n), \qquad (D.1)$$

where $H(p| X_n = n)$ is the posterior distribution of p. Lindley (2008) assumes a uniform distribution for H, and hence the posterior density f of p given $X_n = n$ equals

$$f(p| X_n = n) = c \, P(X_n = n|p) \, f(p) = c \, p^n \cdot 1 = (n+1) \, p^n,$$

where c is a constant such that the integral over this density equals one. Hence (D.1) equals

$$\int_{[0,1]} p^m \, (n+1) \, p^n \, dp = (n+1)/(m+n+1),$$

as presented by Lindley (2008). We see that if m is equal to one and n is large, this probability is close to one – i.e. the probability that the next swan is black is negligible – but if m is large the probability (D.1) is close to zero, i.e. the probability of at least one black swan in the large sample of size m is close to one.

(b) The second part of this appendix shows the computational result of the black swan example presented in Sections 1.7 and 3.4. Here the prior probabilities give mass 0.2 and 0.8 to the p values 1.0 and 0.99, respectively.

The task is again to compute $P(Y_m = m| X_n = n)$. Following arguments as above, we find

$$P(Y_m = m| X_n = n) = P(Y_m = m| X_n = n, p = 1) P(p = 1| X_n = n)$$
$$+ P(Y_m = m| X_n = n, p = 0.99) P(p = 0.99| X_n = n)$$
$$= 1 \cdot P(p = 1| X_n = n) + 0.99^m P(p = 0.99| X_n = n).$$

Now, using Bayes' formula, it is not difficult to see that

$$P(p = 1| X_n = n) = c P(X_n = n|p = 1) P(p = 1) = c \, 1 \cdot 0.2 \text{ and}$$
$$P(p = 0.99| X_n = n) = c P(X_n = n|p = 0.99) P(p = 0.99) = c \, 0.99^n \cdot 0.8,$$
leading to
$$P(Y_m = m| X_n = n) = [0.2/(0.2 + 0.99^n \cdot 0.8)] + [0.8 \cdot 0.99^{m+n}/(0.2 + 0.99^n \cdot 0.8)].$$

We see that in this case if n is large this probability is close to one, i.e. the probability of at least one black swan occurring is close to zero; this is also the case for large m values, which is in contrast to Lindley's result in a above.

References

Abrahamsen, E. and Aven, T. (2012) Why risk acceptance criteria need to be defined by the authorities and not the industry. *Reliability Engineering and System Safety*, 105, 47–50.

Ackoff, R.L. (1989) From data to wisdom. *Journal of Applied Systems Analysis*, 16, 3–9.

Adams, J. (1995) *Risk*. London: UCL Press.

Adler, M.J. (1986) *A Guidebook to Learning for The Lifelong Pursuit of Wisdom*. New York: Collier Macmillan.

Ale, B.J.M. (2002) Risk assessment practices in The Netherlands. *Safety Science*, 40, 105–126.

Althaus, C.E. (2005) A disciplinary perspective on the epistemological status of risk. *Risk Analysis*, 25(3), 567–88.

Ambrose, F. and Ahern, B. (2008) "Unconventional red teaming", Anticipating Rare Events: Can Acts of Terror, Use of Weapons of Mass Destruction or Other High Profile Acts Be Anticipated? Available at: http://redteamjournal.com/papers/U_White_Paper-Anticipating_Rare_Events_Nov2008rev.pdf (Accessed 10 September 2013).

Anderson, D.R., Sweeney, D.J., and Williams, T.A. (1994) *Introduction to Statistics: Concepts and Applications*. St. Paul, MN: West Group.

Angell, F.J. (1959) *Insurance, Principles and Practices*. New York: The Ronald Press Comp., p. 4.

Anoop, M.B. and Rao, K.B. (2008) Determination of bounds on failure probability in the presence of hybrid uncertainties. *Sadhana* 33: 753–765.

Apostolakis, G.E. (1988) The interpretation of probability in probabilistic safety assessments. *Reliability Engineering and System Safety*, 23, 247–252.

——(1990) The concept of probability in safety assessments of technological systems. *Science*, 250, 1359–1364.

——(2004) How useful is quantitative risk assessment? *Risk Analysis*, 24, 515–520.

Atheam, J.L. (1969) *Risk and Insurance*. New York: Appleton-Century-Crofts, p. 36.

Aven, E. and Aven, T. (2011) On how to understand and express enterprise risk. *International Journal of Business Continuity and Risk Management*, 2(1), 20–34.

——(2014) On the link between risk management and the process of meeting overall strategic performance objectives, in an enterprise context. Paper submitted for possible publication.

Aven, T. (2000) Risk analysis – a tool for expressing and communicating uncertainty. *In Proceedings ESREL 2000*, Edinburgh 15–17 May 2000. Cottam, M.P., Harvey, D.W., Pape, R.P. and Tait, J. (eds). Rotterdam: Balkema Publishers, pp. 21–28.

——(2007) A unified framework for risk and vulnerability analysis and management covering both safety and security. *Reliability Engineering and System Safety*, 92, 745–754.

——(2008) *Risk Analysis*. Chichester: Wiley. Reprinted 2009.

——(2009) Safety is the antonym of risk for some perspectives of risk. *Safety Science*, 47, 925–930.

——(2010a) Some reflections on uncertainty analysis and management. *Reliability Engineering and System Safety*, 95, 195–201.

——(2010b) On the need for restricting the probabilistic analysis in risk assessments to variability. *Risk Analysis*, 30(3), 354–360. With discussion 381–384.

——(2010c) *Misconceptions of Risk*. Chichester: Wiley.

——(2011a) *Quantitative Risk Assessment. The Scientific Platform*. Cambridge: Cambridge University Press.

——(2011b) Selective critique of risk assessments with recommendations for improving methodology and practice. *Reliability Engineering and System Safety*, 96, 509–514.

——(2011c) On some recent definitions and analysis frameworks for risk, vulnerability and resilience. *Risk Analysis*, 31(4), 515–522.

——(2011d) On the interpretations of alternative uncertainty representations in a reliability and risk analysis context. *Reliability Engineering and System Safety*, 96, 353–360.

——(2011e) On different types of uncertainties in the context of the precautionary principle. *Risk Analysis*, 31(10), 1515–1525. With discussion 1538–1542.

——(2011f) On the new ISO guide on risk management terminology. *Reliability Engineering and System Safety*, 96(7), 719–726.

——(2012a) *Foundation of Risk Analysis*. 2nd ed. Chichester: Wiley.

——(2012b) The risk concept – historical and recent development trends. *Reliability Engineering and System Safety*, 99, 33–44.

——(2012c) On the link between risk and exposure, *Reliability Engineering and System Safety*, 106 (2012) 191–199.

——(2012d) On when to base event trees and fault trees on probability models and frequentist probabilities in quantitative risk assessments. *International Journal of Performability Engineering*, 8(3), 311–320.

——(2012e) Foundational issues in risk assessment and management. *Risk Analysis*. 32(10), 1647–1656

——(2013a) On the meaning of the black swan concept in a risk context. *Safety Science*, 57, 44–51

——(2013b) A conceptual framework for linking risk and the elements of the data-information-knowledge-wisdom (DIKW) hierarchy. *Reliability Engineering and System Safety*, 111, 30–36.

——(2013c) Probabilities and background knowledge as a tool to reflect uncertainties in relation to intentional acts. *Reliability Engineering and System Safety*, 119, 229–234.

——(2013d) Practical implications of the new risk perspectives. *Reliability Engineering and System Safety*, 115, 136–145.

——(2013e) On Funtowicz & Ravetz's "decision stake – system uncertainties" structure and recently developed risk perspectives frameworks. *Risk Analysis*, 22(2), 270–280.

——(2013f) On how to deal with deep uncertainties in a risk assessment and management context. *Risk Analysis*, 33(12), 2082–91.

——(2013g) A conceptual foundation for assessing and managing risk, surprises and black swans. Paper presented at *Network Safety Conference*, Toulouse 21–23 November, 2013.

——(2013h) How to define and interpret a probability in a risk and safety setting. Discussion paper *Safety Science*, with general introduction by the Associate Editor Genserik Reniers, 2013; 51(1), 223–231.

——(2014a) On the meaning of the special-cause variation concept used in the quality discourse – and its link to unforeseen and surprising events in risk management. *Reliability Engineering and System Safety*, 126, 81–86.

——(2014b, forthcoming) The concept of antifragility and its implications for the practice of risk analysis. Revised and resubmitted to *Risk analysis*.

——(2014c, forthcoming) Confronting black swans. Paper submitted for possible publication.

Aven, T., Asche F., Lindøe, .P, Toft, A., and Wiencke, H.S. (2010) A framework for decision support on HSE regulations. In Menoni, C. (ed.) *Risks Challenging Publics, Scientists and Governments*. CRC Press.

Aven, T., Baraldi, P., Flage, R. and Zio, E. (2014a) *Uncertainties in Risk Assessments*. Chichester: Wiley.

Aven, T. and Bergman, B. (2012) A conceptualistic pragmatism in a risk assessment context. *International Journal of Performability Engineering*, 8(3), 223–232.

Aven, T., Flage, R., Krohn, B.S. and Røed, W. (2014b, forthcoming) Risk, unforeseen events and surprises (black swans) – some examples from the oil and gas industry. Paper submitted for possible publication.

Aven, T. and Guikema, S. (2011) Whose uncertainty assessments (probability distributions) does a risk assessment report: the analysts' or the experts'? *Reliability Engineering and System Safety*, 96, 1257–1262.

Aven, T. and Hiriart, Y. (2013) Robust optimization in relation to a basic safety investment model with imprecise probabilities. *Safety Science*, 55, 188–194.

Aven, T. and Kristensen, V. (2005) Perspectives on risk – Review and discussion of the basis for establishing a unified and holistic approach. *Reliability Engineering and System Safety*, 90, 1–14.

Aven, T. and Krohn, B.S. (2014) A new perspective on how to understand, assess and manage risk and the unforeseen. *Reliability Engineering and System Safety*, 121, 1–10.

Aven, T. and Renn, O. (2009) On risk defined as an event where the outcome is uncertain. *Journal of Risk Research*, 12, 1–11.

——(2010) *Risk Management and Risk Governance*. Berlin: Springer Verlag.

——(2012) On the risk management and risk governance for petroleum operations in the Barents Sea area. *Risk Analysis*, 32(9), 1561–75.

Aven, T., Renn, O. and Rosa, E. (2011) On the ontological status of the concept of risk. *Safety Science*, 49, 1074–1079.

Aven, T. and Vinnem, J.E. (2005) On the use of risk acceptance criteria in the offshore oil and gas industry. *Reliability Engineering and System Safety*, 90, 15–24.

——(2007) *Risk Managment*. NY: Springer Verlag.

Aven, T. and Zio, E. (2011) Some considerations on the treatment of uncertainties in risk assessment for practical decision-making. *Reliability Engineering and System Safety*, 96, 64–74.

——(2013) Model output uncertainty in risk assessment. *International Journal of Performability Engineering*, 9(5), 475–486.

——(2014, forthcoming) Foundational issues in risk assessment and risk management. *Risk Analysis*.

Baraldi, P. and Zio, E. (2008) A combined Monte Carlo and possibilistic approach to uncertainty propagation in event tree analysis. *Risk Analysis* 28: 1309–1325.

Beck, U. (1992) *Risk Society. Towards a New Modernity*. London: Sage.

Bedford, T. and Cooke, R. (2001) *Probabilistic Risk Analysis*, Cambridge: Cambridge University Press.

Bennett, C.T. (1995) The quantitative failure of human reliability analysis. American Society of Mechanical Engineers, *International Mechanical Engineering Congress and Exhibition*, San Francisco, CA, November 1995. www.osti.gov/bridge/servlets/purl/100254-PBmgTY/webviewable/ (Accessed 9 February 2014).

Bergman, B. (2009) Conceptualistic pragmatism: a framework for Bayesian analysis? *IIE Transactions*, 41, 86–93.

Bergman, B. and Klefsjö, B. (2003) *Quality*. 2nd ed. Lund, Sweden: Studentlitteratur.

Bernardo, J. and Smith, A. (1994) *Bayesian Theory*. New York: Wiley.

Berner, C.L. and Flage, R. (2014) Potential uses and limitations in the use of the NUSAP uncertainty and quality assessment scheme in semi-quantitative risk assessment. ESREL 2014.

Bernstein, P.L. (1996) *Against the Gods: The Remarkable Story of Risk*. New York: John Wiley & Sons.

Bersimis, S., Psarakis, S. and Panaretos, J. (2007) Multivariate Statistical Process Control Charts: An Overview. *Quality and Reliability Engineering International*, 23, 517–543.

Bjerga, T. and Aven, T. (2014, forthcoming) Adaptive risk management using the new risk perspectives – an example from the oil and gas industry. Paper submitted for possible publication.

Bjerga, T., Aven, T. and Zio, E. (2014) An illustration of the use of an approach for treating model uncertainties in risk assessment. *Reliability Engineering and System Safety*, 125, 46–53.

Bolstad, M.W. (2007) *Introduction to Bayesian Statistics*. 2nd ed. Hoboken, NY: Wiley.

Borel, E. (1950) *Elements de la theorie des probabilities*. Paris: Albin.

Brown, G.G. and Cox, Jr., L.A. (2011) How probabilistic risk assessment can mislead terrorism risk analysts. *Risk Analysis*, 31(2), 196–204.

Cabinet Office (2002) *Risk: Improving government's capability to handle risk and uncertainty*. Strategy unit report. UK.

Campbell, S. (2005) Determining overall risk. *Journal of Risk Research*, 8, 569–581.

Campbell, S. and Currie, G. (2006) Against Beck: In defence of risk analysis. *Philosophy of the Social Sciences*, 36(2), 149–72.

Carnap, R. (1922) *Der logische Aufbau der Welt*. Berlin.

——(1929) *Abriss der Logistik*. Wien.

Chakhunashvili, A. and Bergman, B. (2007) In weak statistical control? *International Journal of Six Sigma and Competitive Advantage*, 3(1), 91–102.

Chapman, J. (2005) Predicting technological disasters: mission impossible? *Disaster Prevention and Management*, 14(3), 343–52.

Chapman, J.A. and Ferfolja, T. (2001) Fatal flaws: the acquisition of imperfect mental models and their use in hazardous situations. *Journal of Intellectual Capital*, 2(4), 398–409.

Chermack, T.J. (2011) *Scenario Planning in Organizations*. San Francisco: BK Publishers.

Chiasson, P. (2005) Abduction as an aspect of retroduction. *Semiotica*. 2005(153), 223–242..

Clemen, R.T. and Winkler, R.L. (1999) Combining probability distributions from experts in risk analysis. *Risk Analysis*, 19, 187–203.

Cleveland, H. (1982) Information as resource. *The Futurist*, December, 34–39.

Cooke, R. (1991) *Experts in Uncertainty*. Oxford: Oxford University Press.

Cooke, R.M. (1986) Conceptual fallacies in subjective probability. *Topoi* 5, 21–7.

——(2004) The anatomy of the squizzel: The role of operational definitions in representing uncertainty. *Reliability Engineering and System Safety*, 85, 313–319.

Copeland, T.E. and J.F. Weston (1988) *Finance Theory and Corporate Policy*. 3rd ed. Addison-Wesley Publishing Company.

Cottman, R.J. (1993) *Total Engineering Quality Management*. New York: Marcel Dekker.

Courtney, H. (2001) *20/20 Foresight: Crafting strategy in an uncertain world*. Boston: Harvard Business School Press.

Cowell, R.G., Dawid, A.P., Lauritzen, S.L. and Spiegelhalter, D.J. (1999) *Probabilistic Networks and Expert Systems*. New York: Springer.

Cox, L.A.T. (2012) Confronting deep uncertainties in risk analysis. *Risk Analysis*, 32, 1607–1629.

Cox, T. (2011) Clarifying types of uncertainty: When are models accurate, and uncertainties small? *Risk Analysis*, 31, 1530–33.

Crowe, R.M. and Horn, R.C. (1967) The meaning of risk. *The Journal of Risk and Insurance* 34(3), 459–474.

Cullen, W.D. (1990) *The Public Inquiry into the Piper Alpha Disaster*. London: Department of Energy.

Cumming, R.B. (1981) Is risk assessment a science? *Risk Analysis*, 1, 1–3.

Cutler, A. (1997) Deming's vision applied to probabilistic risk analysis. Presented at the *2nd Edinburgh Conference on Risk: Analysis, Assessment and Management*, Edinburgh, UK, September 1997.

DCPEP (2013) National risk analysis. The Norwegian Directorate for Civil Protection and Emergency Planning. www.dsb.no/Global/Publikasjoner/2013/Tema/NRB_2013_english.pdf (Accessed 8 February 2014).

Deepwater (2013) The National Commission on the Deepwater Horizon Oil Spill and Offshore Drilling. www.oilspillcommission.gov/final-report (Accessed 28 March 2013).

de Finetti, B. (1930) Fondamenti Logici del Ragionamento Probabilistico. *Bollettino dell'Unione Matematica Italiana*, 5, 1–3.

——(1974) *Theory of Probability*. NY: Wiley.

Dekker, S. (2011) *Drift into Failure: From Hunting Broken Components to Understanding Complex Systems*. Aldershot, Hampshire: Ashgate Publishing.

Dekker, S.W.A. and Nyce, J.M. (2004) How can ergonomics influence design? Moving from research findings to future systems. *Ergonomics*, 47(15), 1624–39.

de Laplace, P.S. (1812) *Théorie analytique des probabilités*. Paris: Courcier Imprimeur.

Deming, W.E. (2000) *The New Economics*. 2nd ed. Cambridge, MA: MIT CAES.

de Moivre, A. (1711) De Mensura Sortis. *Philosophical Transactions*, 27, 213–64.

Dempster, A.P. (1967) Upper and lower probabilities induced by a multivalued mapping. *Annals of Math. Statistics* 38: 325–339.

Douglas, M. and Wildavsky, A. (1982) *Risk and Culture: The Selection of Technological and Environmental Dangers*. Berkeley, CA: University of California Press.

Dubois, D. (2006) Possibility theory and statistical reasoning. *Computational Statistics and Data Analysis*, 51, 47–69.

——(2010) Representation, propagation and decision issues in risk analysis under incomplete probabilistic information. *Risk Analysis*, 30, 361–368.

Dubois, D. and Prade, H. (1989) Fuzzy sets, probability and measurement. *European Journal. of Operations Research* 40: 135–154.

Eidesen, K. and Aven, T. (2010) Uncertainty assessments in a semi-quantitative risk analysis, with application to health care. In: Bris, R., Guedes Soares, C. and Martorell, S. *Reliability, Risk and Safety: Theory and Applications. Proceedings of the European Safety and Reliability Conference 2009* (ESREL 2009), Prague, Czech Republic, 7–10 September 2009, pp. 1803–1808.

Eliot, T.S. (1934) *The Rock*. London, Faber and Faber.

Ferson, S. and Ginzburg, L.R. (1996) Different methods are needed to propagate ignorance and variability. *Reliability Engineering and System Safety*, 54, 133–144.

Financial Post (2013) http://business.financialpost.com/2013/03/14/halliburton-worker-weeps-as-he-admits-he-overlooked-warning-of-blast-that-set-off-americas-biggest-oil-spill-in-gulf/?__lsa=42e0-28bb (Accessed 20 December 2013).

Fischhoff, B., Lichtenstein, S., Slovic, P., Derby, S. and Keeney, R. (1981) *Acceptable Risk*. New York: Cambridge University Press.

Flage, R. and Aven, T. (2009) Expressing and communicating uncertainty in relation to quantitative risk analysis (QRA). *Reliability and Risk Analysis: Theory and Applications*, 2(13), 9–18.

Flanders, W.D., Lally, C.A., Zhu, B-P., Henley, S.J. and Thun, M.J. (2003) Lung cancer mortality in relation to age, duration of smoking, and daily cigarette consumption. *Cancer Research*, 63, 6556–6562.

Flood, R.L. (1999) *Rethinking the Fifth Discipline: Learning within the Unknowable*. London: Routledge Publishing.

Franklin, J. (2001) Resurrecting logical probability. *Erkenntnis*, 55(2), 277–305.

French, S. and Insua, D.R. (2000) *Statistical decision theory*. London: Arnold.

Frické, M. (2009) The knowledge pyramid: A critique of the DIKW hierarchy. *Journal of Information Science*, 35(2), 131–142.

Funtowicz, S.O. and Ravetz, J.R. (1990). *Uncertainty and Quality in Science for Policy*. Dordrecht: Kluwer Academic Publishers.

——(1993). Science for the post-normal age. *Futures*, 25, 735–755.

Furlong, R.B (1984) Clausewitz and Modern War Gaming: Losing can be better than winning. *Air University Review*, July–August 1984.

Galavotti, M.C. (2000) The modern epistemic interpretations of probability: logicism and subjectivism. In *Handbook of the History of Logic*. Volume 10:

Inductive Logic. Volume editors: D.M. Gabbay, S. Hartmann and J. Woods. General editors: D.M. Gabbay, P. Thagard and J. Woods. London: Elsevier.

Giddens, A. (1999) *Runaway World: How Globalisation is Reshaping Our Lives*. London: Profile Books, pp. 21–22, 35.

Gigerenzer, G., Swijtink, Z., Porter, T., Daston, L., Beattie, J. and Kruger, L. (1989) *The Empire of Chance: How Probability Changed Science and Everyday Life*. Cambridge: Cambridge University Press.

Gillies, D. (2000) *Philosophical Theories of Probability*. London: Routledge.

Goerigk, M. and Schöbel, A. (2011) A scenario-based approach for robust linear optimization. Marchetti-Spaccamela, A., and Segal, M. (eds.) *Lecture Notes in Computer Science 6595*. Berlin Heidelberg: Springer-Verlag, pp. 139–150.

Good, I.J. (1951) Rational decisions. *Journal of the Royal Statistical Society*, Ser. B, 14, 107–14.

Gross, M. (2010) *Ignorance and Surprises*. London: MIT Press.

Hajek, A. (2001) Probability, logic and probability logic. In Goble, L. (ed.) *The Blackwell Companion to Logic*. Blackwell, pp. 362–384.

——(2007) The reference class problem is your problem too. *Synthese*, 56, 563–585.

Hammond, P. (2009) Adapting to the entirely unpredictable: Black swans, fat tails, aberrant events, and hubristic models. *The University of Warwick Bulletin of the Economics Research Institute*, 2009/10, 1, November.

Hanley, N. and Spash, C.L. (1993) *Cost-benefit Analysis and the Environment*. Aldershot, England: E. Elgar.

Hansson, S.O. (2002) Uncertainties in the knowledge society. *International Social Science Journal*, 54(171), 39–46.

Hansson, S.O. (2013) Defining pseudoscience and science. In Pigliucci, M. and Boudry, M. *Philosophy of pseudoscience*. Chicago: University of Chicago Press, pp. 61–77.

Hansson, S.O. and Aven, T. (2014) Is risk analysis scientific? *Risk Analysis*. In Press.

Hardy, C.O. (1923) *Risk and Risk Bearing*. Chicago: University of Chicago, p. 1.

Harrowitz, N. (1983) The Body of the Detective Model: Charles S. Peirce and Edgar Allan Poe. In Eco, U. and Sebeok, T.A. (eds.) *The Sign of Three: Dupin, Holmes, Peirce*. Indiana: Indiana University Press.

Haynes, J. (1895) Risk as an economic factor. *The Quarterly Journal of Economics*, 9(4), (July, 1895), 409.

Hey, J. (2012) The data, information, knowledge, wisdom chain: The metaphorical link. http://en.wikipedia.org/wiki/Jonathan_Hey (Accessed 13 January 2012).

HT Politics (2011) *Spill commission calls for much tougher oversight of oil drilling* (January 11, 2011). http://politics.heraldtribune.com/2011/01/11/spill-commission-calls-for-much-tougher-oversight-of-oil-drilling/ (Accessed 1 May 2014).

Hoffman, F.O. and Kaplan, S. (1999) Beyond the domain of direct observation: how to specify a probability distribution that represents the 'state of knowledge' about uncertain inputs. *Risk Analysis*, 19, 131–134.

Hollnagel, E., Woods, D. and Leveson, N. (2006) *Resilience Engineering: Concepts and Precepts*. UK: Ashgate.

Hopkins, A. (2000) *Lessons from Longford: The Esso Gas Plant Explosion*. Sydney: CCH Australia Ltd.

——(2014) Issues in safety science. *Safety Science*. 67, 6–14.

Hopkins, A. and Hale, A. (2002). Issues in the regulation. In Kirwan, B., Hale, A. and Hopkins, A., (eds.) *Changing Regulation Controlling Risks in Society.* Amsterdam: Pergamon.

Hudson, R.G. (1994) Reliability, pragmatic and epistemic. *Erkenntnis*, 40(1), 71–86.

IRGC, International Risk Governance Council (2005) *White Paper on Risk Governance. Towards an Integrative Approach.* Author: O. Renn with Annexes by P. Graham. Geneva: International Risk Governance Council.

ISO (2009a) *Risk Management – Vocabulary.* Guide 73: 2009.

——(2009b) *Risk Management – Principles and Guidelines.* ISO 31000: 2009.

Jasanoff, S. (1999) The songlines of risk. *Environmental Values.* Special Issue: Risk, 8(2), 135–52.

Jeffrey, R. (2004) *Subjective Probability: The Real Thing.* Cambridge: Cambridge University Press.

Kaplan,S. (1988) Will the real probability please stand up? *Reliability Engineering and System Safety*, 23(4), 285–292.

——(2000) Combining probability distributions from experts in risk analysis. *Risk Analysis*, 20(2), 155–56.

Kaplan, S. and Garrick, B.J. (1981) On the quantitative definition of risk. Risk Analysis 1, 11–27.

Kaplan, S., Visnepolschi, S., Zlotin, B. and Zusman, A. (1999) *New Tools for Failure and Risk Analysis: Anticipatory Failure Determination (AFD) and the Theory of Scenario Structuring.* Southfield, MI: Ideation International Inc.

Karvetski, C.W. and Lambert, J.H. (2012) Evaluating deep uncertainties in strategic priority-setting with an application to facility energy investments. *Systems Engineering*, 15(4), 483–493.

Kasperson, R.E. (2008) Coping with Deep Uncertainty: Challenges for environmental assessment and decision making. In Smitson, A.M. and Bammer, G. (ed.) *Uncertainty and Risk: Multidisciplinary Perspectives.* London: Earthscan.

Kay, N.M. (1984) *The Emergent Firm: Knowledge, Ignorance and Surprise.* London: Macmillan.

Kennedy, M.C. and O'Hagan, A. (2001) Bayesian calibration of computer models. *Journal of the Royal Statistical Society*, Series B (Statistical Methodology), 63(3), 425–464.

Keynes, J. (1921) *Treatise on Probability.* London.

Khorsandi, J. and Aven, T. (2014, forthcoming) Understanding risk: A risk perspective suitable for organizations seeking high reliability. *Journal of Risk Research.* In press.

Khorsandi, J., Aven, T. and Vinnem, J.E. (2012) A review and discussion of the Norwegian offshore safety regulations regime for risk assessments. In *Proceedings PSAM11-ESREL 2012 Conference* Helsinki, June 25–29. Red Hook, NY: Curran Associates, Inc ISBN 9781622764365, pp. 5702–12.

Kirchsteiger, C. (2002) Preface: International workshop on promotion of technical harmonisation on risk-based decision-making. *Safety Science*, 40, 1–15.

Kloprogge, P., van der Sluijs, J. and Petersen, A. (2005) *A Method for the Analysis of Assumptions in Assessments.* Bilthoven, The Netherlands: Netherlands Environmental Assessment Agency (MNP).

Kloprogge, P., van der Sluijs, J.P. and Petersen, A.C. (2011) A method for the analysis of assumptions in model-based environmental assessments. *Environmental Modelling and Software*, 26, 289–301.

Knight, F.H. (1921) *Risk, Uncertainty, and Profit*. Boston: Houghton Mifflin Co., p. 233.

Knupp, P. (2002) *Verification of Computer Codes in Computational Science and Engineering*. Boca Raton, FL: Chapman & Hall/CRC.

Kunreuther, H. and Useem, M. (2010) *Learning from Catastrophes. Strategies for Reaction and Response*. Upper Saddle River, New Jersey: Wharton School Publishing.

Laes, E., Meskens, G., and van der Sluijs, J.P. (2011) On the contribution of external cost calculations to energy system governance: The case of a potential large-scale nuclear accident. *Energy Policy*, 39, 5664–5673.

Latour, B. (2005) *Reassembling the Social: An Introduction to Actor Network Theory*. Oxford: Oxford University Press.

Le Coze, J-C. (2013) Outlines of a sensitising model for industrial safety assessment. *Safety Science*, 51, 187–201.

Leistad, G.H. and Bradley, A.R. (2009) Is the focus too low on issues that have a potential that can lead to a major incident? SPE 123861. Paper presented at *SPE Offshore Europe Oil and Gas Conference* Aberdeen 8–11 Sept 2009.

Lemos, N. (2007) *An Introduction to the Theory of Knowledge*. Cambridge: Cambridge University Press.

Lempert, R.J., Light, P.C., Pritchett, L., Treverton, G.F., Groves, D.G., Popper, S.W., Bankes, S.C., Collins, M.T., Nakicenovic, N. and Sarewitz, D. (2004) *Shaping the Next One Hundred Years: New Methods for Quantitative, Long-Term Policy Analysis*. Santa Monica, CA: RAND.

Leveson, N. (2011) *Engineering a Safer World*. Cambridge: The MIT Press.

Levy, H. and Sarnat, M. (1994) *Capital Investment and Financial Decisions*. 5th ed. New York: Prentice Hall.

Lewis, C.I. (1929) *Mind and the World Order: Outline of a Theory of Knowledge*. New York, NY: Dover Publications.

Lindaas, O.A. and Pettersen, K. (2013) Risk communication and black swans – dealing with uncertainty by categorization. Paper resented at *11th International Conference on Structural Safety & Reliability*, Columbia University, New York, June 16–20, 2013.

Lindley, D.V. (1985) *Making Decisions*. New York: Wiley.

——(2000) The philosophy of statistics. *The Statistician*, 49, 293–337. With discussions.

——(2006) *Understanding Uncertainty*. Hoboken, NJ: Wiley.

——(2008) The Black Swan: the impact of the highly improbable. Reviews. Significance, p. 42. March 2008.

Linkov, I., Satterstrom, F., Kiker, G., Batchelor, C., Bridges, T. and Ferguson, E. (2006) From comparative risk assessment to multi-criteria decision analysis and adaptive management: Recent developments and applications. *Environment International*, 32, 1072–1093.

Littlewood, B. and Strigini, L. (2004) Redundancy and diversity in security. *Proceedings of the 9th European Symposium on Research in Computer Security*, Sophia Antipolis, France, September, pp. 423–438. Springer-Verlag, Lecture Notes in Computer Science 3193, 2004.

Littlewood, B., Brocklehurst, S., Fenton, N., Mellor, P., Page, S., Wright, D., Dobson, J., McDermid, J. and Gollmann, D. (1993) Towards operational measures of computer security, *Journal of Computer Security*, 2, 211–229.

Lowrance, W. (1976) *Of Acceptable Risk – Science and the Determination of Safety*. Los Altos, CA: William Kaufmann Inc.

Lupton, D. (1999) *Risk*. London: Routledge, p. 5.

Magee, J.H. (1961) *General Insurance*. 6th ed. Homewood, Illinois: Richard D. Irwin.

Magnussen, K., Kroslid, D. and Bergman, B. (2003) *Six Sigma, the Pragmatic Approach*. 2nd ed. Lund, Sweden: Studentlitteratur.

Majeske, K.D. and Hammett, P.C. (2003) Identifying sources of variation in sheet metal stamping. *The International Journal of Flexible Manufacturing Systems*, 15, 5–18.

Martz, H.F. and Waller, R.A. (1988) On the meaning of probability. *Reliability Engineering and System Safety*, 23(4), 299–304.

Masys, A.J. (2012) Black swans to grey swans: revealing the uncertainty. *Disaster Prevention and Management*, 21(3) 320–335.

McFarland, J.M. (2008) *Uncertainty Analysis for Computer Simulations through Validation and Calibration*. PhD Thesis, Vanderbilt University.

Meeker, W.O. and Escobar, L.A. (1998) *Statistical Methods for Reliability Data*. Wiley: N.Y.

Mehr, R.I. and Cammack, E. (1953) *Principles of Insurance*. 3rd ed. Homewood, Illinois: Richard D. Irwin.

Mellor, D.H. (2005) *Probability: A Philosophical Introduction*. London: Routledge.

Merkelsen, H. (2011) The constitutive element of probabilistic agency in risk: a semantic analysis of risk, danger, chance and hazard. *Journal of Risk Research*, 14(7), 881–897.

Merton, R.K. (1973) Science and technology in a democratic order. *J Legal Pol Sociol*. 1942; 1, 115–126. Reprinted as The normative structure of science. In Merton, R.K. *The Sociology of Science. Theoretical and Empirical Investigations*. Chicago: University of Chicago Press 1973, pp. 267–278.

Meyer, T. and Reniers, G. (2013) *Engineering Risk Management*. Berlin: De Gruyter Graduate.

Mosleh, A. and Bier, V. (1996) Uncertainty about probability: a reconciliation with the subjectivist viewpoint. *Systems, Man and Cybernetics, Part A: Systems and Humans*, 26(3), 303–310.

Mowbray, A.H. and Blanchard, R.H. (1961) *Insurance*. 5th ed. New York: McGraw-Hill.

Munsterhjelm-Ahumada, K. (2012) Health Authorities Now Admit Severe Side Effects of Vaccination: Swine Flu, Pandemrix and Narcolepsy, Orthomolecular Medicine News Service, March 20. http://orthomolecular.org/resources/omns/v08n10.shtml (Accessed 11 January 2014).

Myers, N. (1993) Biodiversity and the Precautionary Principle. Ambio, 22, No. 2/3, Biodiversity: Ecology, Economics, Policy (May, 1993), pp. 74–79. Published by Allen Press on behalf of the Royal Swedish Academy of Sciences.

North, W. (2010) Probability theory and consistent reasoning, commentary. *Risk Analysis*, 30(3), 377–380.

NUREG (2009) US Nuclear Regulatory Commission. *Guidance on the Treatment of Uncertainties Associated with PRAs in Risk-Informed Decision Making* (NUREG-1855). Washington DC, 2009.

Oakland, J.S. (2003) *Statistical Process Control.* 5th ed. Oxford: Butterworth-Heinemann.

Oberkampf, W. and Trucano, T. (2002) *Verification and Validation in Computational Fluid Dynamics.* Technical Report SAND2002-0529. Albuquerque, New Mexico: Sandia National Laboratories.

OECD (2009) *OECD studies in risk management: Innovation in country risk management.* Organisation for Economic Co-operation and Development.

Offshore (2011) PSA seeks answers on Gullfaks leak. www.offshore-mag.com/articles/2011/03/psa-seeks-answers.html (Accessed 8 February 2014).

O'Hagan, A. and Oakley, J.E. (2004) Probability is perfect, but we can't elicit it perfectly. *Reliability Engineering and System Safety*, 85, 239–248.

Östberg, G. (1988) On the meaning of probability in the context of probabilistic safety assessment. *Reliability Engineering and System Safety*, 23(4), 305–308.

Oxford English Dictionary (2014) www.oed.com (Accessed 1 February 2014).

Parry, G.W. (1988) On the meaning of probability in probabilistic safety assessment *Reliability Engineering and System Safety*, 23(4), 309–314.

Paté-Cornell, M. (1996) Uncertainties in risk analysis: Six levels of treatment, *Reliability Engineering and System Safety*, 54(2–3), 95–111.

Paté-Cornell, M.E. (2012) On black swans and perfect storms: risk analysis and management when statistics are not enough. *Risk Analysis*, 32(11), 1823–1833.

Petitti, D.B. (2000) *Meta-Analysis, Decision Analysis, and Cost-Effectiveness Analysis.* Oxford: Oxford University Press.

Pettersen, K.A. (2013) Acknowledging the role of abductive thinking: A way out of proceduralization for safety management and oversight? In Bieder, C. and Bourrier, M. (eds.) *Trapping Safety into Rules. How Desirable or Avoidable is Proceduralization?* Surrey, UK: Ashgate.

Pfeffer, I. (1956) *Insurance and Economic Theory.* Homewood, Illinois: Richard D. Irwin Inc., p. 42.

Popper, K. (1962) *Conjectures and Refutations. The Growth of Scientific Knowledge.* New York: Basic Books.

PSA-N (2013a) *Principles for barrier management in the petroleum industry.* Norwegian Petroleum Safety Authority (PSA-N), 29/1–2013. www.ptil.no/getfile.php/PDF/Barrierenotatet%202013%20engelsk%20april.pdf (Accessed 9 February 2014).

PSA-N (2013b) *Assessments and recommendations after Deepwater Horizon.* Petroleum Safety Authority Norway (PSA-N). www.ptil.no./well-integrity/assessments-and-recommendations-after-deepwater-horizon-article7890-900.html (Accessed 1 May 2014).

PST (2013) Norwegian Police Security Service (PST). www.politi.no/vedlegg/rapport/Vedlegg_882.pdf (Accessed 1 February 2014).

Ramsey, F. (1931) Truth and probability. In *Foundations of Mathematics and Other Logical Essays.* London: Routledge and Kegan Paul, 156–198.

Ranger, N., Reeder, T. and Lowe, J. (2013) Addressing 'deep' uncertainty over long-term climate in major infrastructure projects: four innovations of the Thames Estuary 2100 Project. *EURO J Decis Process*, 1:233–262

Reason, J. (2004) Beyond the organisational accident: the need for 'error wisdom' on the Frontline. *Quality and Safety in Healthcare*, 13(2), 28–33.

Renn, O. (2008) *Risk Governance: Coping with Uncertainty in a Complex World*. London: Earthscan.

Renn, O. and Klinke, A (2002) A new approach to risk evaluation and management: risk-based precaution-based and discourse-based strategies. *Risk Analysis*, 22(6):1071-94.

Riegel, R. and Miller, J.S. (1966) *Insurance Principles and Practices*. 5th ed. Englewood Cliffs, NJ: Prentice-Hall, Inc., p. 20.

Rochlin, G.I. (1999) Safe operation as a social construct. *Ergonomics*, 42(11), 1549–60.

Rosa, E.A. (1998) Metatheoretical foundations for post-normal risk. *Journal of Risk Research* 1, 15–44.

——(2003) The logical structure of the social amplification of risk framework (SARF): Metatheoretical foundation and policy implications. In Pidgeon, N., Kaspersen, R.E. and Slovic, P. (eds.) *The Social Amplification of Risk*. Cambridge: Cambridge University Press.

Rowe, W.D. (1977) *An Anatomy of Risk*. New York: John Wiley & Sons, p. 24.

Rowley, J. (2006) Where is the wisdom that we have lost in knowledge? *Journal of Documentation*, 62(2), 251–270.

——(2007) The wisdom hierarchy: representations of the DIKW hierarchy. *Journal of Information Science*, 33(2), 163–180.

Røed, W., Vinnem, J.E. and Nistov, A. (2012) Causes and contributing factors to hydrocarbon leaks on Norwegian offshore installations. Paper presented at the *SPE/APPEA International Conference on Health, Safety, and Environment in Oil and Gas Exploration and Production*, Perth, Australia, 11–13 September 2012. SPE 156846.

Sande T. (2013) Misunderstood risk (in Norwegian). Misforstått risiko. Stavanger Aftenblad, 1 October, p. 20.

Savage, L. (1956) *Foundations of Statistics*, New York: John Wiley & Sons.

——(1962) On the foundations of statistical inference: Discussion. *Journal of the American Statistical Association*, 57(298), 307–326.

Scholz, R.W., Blumer, Y.B. and Brand, F.S. (2012) Risk, vulnerability, robustness, and resilience from a decision-theoretic perspective. *Journal of Risk Research*, 15(3), 313–330.

Senge, P. (1990) *The Fifth Discipline: The Art and Practice of the Learning Organization*. New York, NY: Doubleday Currency.

SEP (2009) Stanford Encyclopedia Philosophy (Interpretations of probability. http://plato.stanford.edu/entries/probability-interpret/ (Accessed 9 February 2014).

Shafer,G. (1976) *A Mathematical Theory of Evidence*. Princeton: Princeton University Press.

——(1990) Perspectives on the theory and practice of belief functions. *International Journal of Approximate Reasoning*. 4, 323–362.

Shewhart, W.A. (1931) *Economic Control of Quality of Manufactured Product*. New York: Van Nostrand.

Shewhart, W.A. (1939) *Statistical Method from the Viewpoint of Quality Control*. Washington DC: Dover Publications.

Shrader-Frechette, K.S. (1991) *Risk and Rationality*. Berkeley, Los Angeles and Oxford: University of California Press.

Singpurwalla, N.D. (2006) *Reliability and risk – A Bayesian Perspective*. Chichester: John Wiley & Sons, Inc.

Solberg, Ø. and Njå, O. (2012) Reflections on the ontological status of risk. *Journal of Risk Research*. 15(9), 1201–1215.

Spiegelhalter, D.J. and Riesch, H. (2014) Don't know, can't know: embracing deeper uncertainties when analysing risks. *Phil. Trans. R. Soc. A*, 369, 4730–4750.

Starossek, U.P.E. and Haberland, M. (2010) Disproportionate collapse: Terminology and procedures. *Journal of Performance of Constructed Facilities*, 24(6), 519–528.

Sternberg, R.J. (1999) *Handbook of Creative Thinking*. Cambridge University Press, Cambridge.

Taleb, N.N. (2007) *The Black Swan: The Impact of the Highly Improbable*. London: Penguin.

——(2010) *The Black Swan: The Impact of the Highly Improbable*. 2nd ed. London: Penguin.

——(2012) *Anti Fragile*. London: Penguin.

——(2013) www.fooledbyrandomness.com/DerivTBS.htm (Accessed 19 December 2013).

Thompson, K.M., Deisler. Jr., P.H. and Schwing, R.C. (2005) Interdisciplinary vision: The first 25 years of the Society for Risk Analysis (SRA), 1980–2005. *Risk Analysis*, 25, 1333–86.

Tickner, J. and Kriebel, D. (2006) The role of science and precaution in environmental and public health policy. In Fisher, E., Jones, J. and von Schomberg, R. (eds.) *Implementing the Precautionary Principle*. Northampton, MA, USA: Edward Elgar Publishing.

Turner, B. and Pidgeon, N. (1997) *Man-made Disasters*. 2nd ed. London: Butterworth-Heinemann.

UCB (2010) *Final report on the investigation of the Macondo well blowout*. University of California Berkeley Report, p. 82. http://ccrm.berkeley.edu/pdfs_papers/bea_pdfs/dhsgfinalreport-march2011-tag.pdf (Accessed 9 February 2014).

UK Gov (2014) *Terrorism and national emergencies. Terrorism threat levels*. UK Government. www.gov.uk/terrorism-national-emergency (Accessed 1 February 2014).

Van der Merwe, L. (2008) Scenario-based strategy in practice: a framework. *Advances in Developing Human Resources*, 10(2), 216–39.

van der Sluijs, J., Craye, M., Futowicz, S., Kloprogge, P., Ravetz, J. and Risbey, J. (2005a) Combining quantitative and qualitative measures of uncertainty in model-based environmental assessment. *Risk Analysis*, 25(2), 481–492.

——(2005b) Experiences with the NUSAP system for multidimensional uncertainty assessment in model based foresight studies. *Water Science and Technology*, 52(6), 133–144.

van Lambalgen, M. (1990) The axiomatisation of randomness. *Journal of Symbolic Logic*, 55(3), 1143–1167.

Varian, H.R. (1999) *Intermediate Microeconomics: A Modern Approach*. 5th ed. New York: W.W. Norton & Company.

Vatn, J. (1998) A discussion of the acceptable risk problem. *Reliability Engineering and System Safety*, 61, 11–9.

Vaurio, J.K. (1990) On the meaning of probability and frequency. *Reliability Engineering and System Safety*, 28(1), 121–130.

Veland, H., Amundrud, H. and Aven, T. (2013) Foundational issues in relation to national risk assessment methodologies. *Journal of Reliability and Risk*, 227, 348–358.

Veland, H. and Aven, T. (2013) Risk communication in the light of different risk perspectives. *Reliability Engineering and System Safety*, 110, 34–40.

——(2014, forthcoming) Improving the risk assessments of critical operations to better reflect uncertainties and the unforeseen. Paper submitted for possible publication.

Verma, M. and Verter, V. (2007) Railroad transportation of dangerous goods: Population exposure to airborne toxins. *Computers and Operations Research* 34, 1287–1303.

Vinnem, J.E. (2007) *Offshore Risk Assessment: Principles, Modelling and Applications of QRA Studies*. 2nd ed. NY: Springer Verlag.

——(2010) Risk analysis and risk acceptance criteria in the planning processes of hazardous facilities – a case of an LNG plant in an urban area. *Reliability Engineering and System Safety*, 95(6), 662–670.

——(2012) On the analysis of hydrocarbon leaks in the Norwegian offshore industry. *Journal of Loss Prevention in the Process Industries*, 25(4), 709–717.

Von Neumann, J. and Morgenstern, O. (1944) *Theory of Games and Economic Behaviour*. Princeton, NJ: Princeton University Press.

Vose, D. (2008) Risk analysis: A quantitative guide. 3rd ed. Chichester, Wiley.

Walker, W.E., Harremoës, P., Rotmans, J., van der Sluijs, J.P., van Asselt, M.B.A., Janssen, P. and Krayer von Krauss, M.P. (2003) Defining uncertainty: a conceptual basis for uncertainty management in model-based decision support. *Integrated Assessment*, 4(1), 5–17.

Walker, W.E., Haasnoot, M. and Kwakkel, J.H. (2013) Adapt or Perish: A Review of Planning Approaches for Adaptation under Deep Uncertainty. *Sustainability 5*, 955–979.

Wallace, R. (2012) Fukushima crushed by 'myth', says panel. *The Australian*, 24 July. Retrieved 30 July 2012.

Walley, P. (1991) *Statistical Reasoning with Imprecise Probabilities*. NY: Chapman and Hall.

Watson, S.R. (1994) The meaning of probability in probabilistic safety analysis. *Reliability Engineering and System Safety*, 45(3), 261–269.

——(1995) Response to Yellman & Murray's (1995) comment on "The meaning of probability in probabilistic risk analysis". *Reliability Engineering and System Safety*, 49(2), 207–209.

Watson, S.R. and Buede, D.M. (1987) *Decision Synthesis*. New York: Cambridge University Press.

Weick, K.E. and Sutcliffe, K.M. (2007) *Managing the Unexpected: Resilient Performance in an Age of Uncertainty*. 2nd ed. San Francisco, CA: John Wiley and Sons Inc.

Weick, K.E., Sutcliffe, K.M. and Obstfeld, D. (1999) Organizing for high reliability: processes of collective mindfulness. *Research in Organizational Behavior*, 2, 13–81.

Weinberg, A.M. (1974) Science and trans-science. *Minerva*, 10(2), 209–222.

——(1981) Reflections on risk assessment. *Risk Analysis*, 1, 5–7.

Wharton, F. (1992) Risk management: Basic concepts and general principles. In Ansell, J. and Wharton, F. (eds) *Risk: Analysis, Assessment and Management*. Chichester: John Wiley and Sons, p. 4.

Wikipedia (2013) http://en.wikipedia.org/wiki/Common_cause_and_special_cause_ (statistics) (Accessed 10 May 2013).

Willett, A.H. (1901) *The Economic Theory of Risk and Insurance*. Philadelphia: The University of Pennsylvania Press, 1951 [originally published in 1901], p. 6.

Willis, H.H. (2007) Guiding resource allocations based on terrorism risk. *Risk Analysis* 27(3), 597–606.

Winkler, R.L. (1996) Uncertainty in probabilistic risk assessment. Reliability Engineering and System Safety 85, 127–132.

Wintle, B. and Lindenmayer, D. (2008) Adaptive risk management for certifiably sustainable forestry. *Forest Ecology and Management*, 256, 1311–1319.

Woodall, W.H. (2000) Controversies and contradictions in statistical process control. *Journal of Quality Technology*, 32(4), 341–350.

Wood, O.G. Jr. (1964) Evolution of the concept of risk. *Journal of Risk and Insurance* 31(1), 83–91.

Wynne, B. (1992) Risk and social learning: Reification to engagement. In Krimsky, S. and Golding, D. (eds) *Social Theories of Risk*. Westport CT: Praeger, pp. 275–97.

Yamaguchi, M. (2012) Fukushima Nuclear Disaster Report: Plant operators Tokyo Electric and Government still stumbling. Associated Press. *The Huffington Post*, 23 July. Retrieved 29 July 2012.

Yamaguchi, N., Kobayashi, Y.M. and Utsunomiya, O. (2000) Quantitative relationship between cumulative cigarette consumption and lung cancer mortality in Japan. *International Journal of Epidemiology*, 29(6), 963–968.

Yellman, T.W. and Murray, T.M. (1995) Comment on "The meaning of probability in probabilistic safety analysis". *Reliability Engineering and System Safety*, 49(2), 201–205.

Zeleny, M. (1987) Management support systems: Towards integrated knowledge management. *Human Systems Management*, 7(1), 59–70.

——(2006) Knowledge-information autopoietic cycle: Towards the wisdom systems. *International Journal of Management and Decision Making*, 7(1), 3–18.

Zio, E. (2006) A study of the bootstrap method for estimating the accuracy of artificial neural networks in predicting nuclear transient processes. *IEEE Transactions on Nuclear Science*, 53(3), 1460–1478.

——(2007) *An Introduction to the Basics of Reliability and Risk Analysis*. Singapore: World Scientific Publishing.

Index

Page numbers in **bold** indicate tables and in *italic* indicate figures.